POLICING THE CAMPUS

COUNTERPOINTS

Studies in the Postmodern Theory of Education

Shirley R. Steinberg
General Editor

Vol. 410

The Counterpoints series is part of the Peter Lang Education list.
Every volume is peer reviewed and meets
the highest quality standards for content and production.

PETER LANG
New York • Washington, D.C./Baltimore • Bern
Frankfurt • Berlin • Brussels • Vienna • Oxford

PRAISE FOR

POLICING THE CAMPUS

"Whether you agree or disagree with their conclusions, the essays in this volume are certain to inspire a much-needed dialogue about the effects of police and military presences on college campuses. Everyone interested in the current state of higher education and its role in democratic society is sure to be equal parts intrigued and incensed by what they find within these pages."

—Steven W. Halady, Visiting Assistant Professor of Philosophy, Canisius College

"We live in a time of unprecedented clampdown on student dissent and 'policed pedagogy,' super-surveillance at colleges across the planet. This book, concentrating on realities of repression largely in the U.S., will galvanize researchers elsewhere to investigate and struggle against similar forms of academic fascism engulfing campuses worldwide."

—Bill Templer, independent scholar, Bulgaria

"As one who has had his classes infiltrated over the years by students from right-wing organizations—solely for the purpose of disrupting (and reporting on) my course content, I cannot say enough about the timeliness and importance of this excellent work. Academic freedom has been under attack for many decades and in many ways. Much ground has been lost that needs to be regained. I hope that this book will contribute toward the accomplishment of that goal."

—John C. Alessio, Former SBS Dean, Minnesota State Mankato, and author of *Social Problems and Inequality: Social Responsibility through Progressive Sociology*

"Anthony J. Nocella II and David Gabbard have done it again! *Policing the Campus* minces no words and pulls no punches to focus the scholarly microscope on the repression and suppression of intellectual thought and action. Faculty and students alike will view their educational opportunities and work in a new light through this startling exposé of academic policing."

—Julie Andrzejewski, Professor, Department of Human Relations and Multicultural Education, St. Cloud State University

"In *Policing the Campus*, Anthony J. Nocella II and David Gabbard make a critical and radical intervention that challenges the policing and surveillance of higher education. If the influence of military, corporations, and law enforcement—and the creation of a police state on every campus—go unchecked, it will be impossible for democratic and free education to flourish."

—Sarat Colling, Co-founder of *Political Media Review*

"In the ever-changing climate of higher education, the policies related to this system still reflect a cultural climate that does little to recognize the diverse community which higher education has become. The book, *Policing the Campus*, pushes us to look at the climate of higher education and the response needed to move both our society and educational instruction to reflect and act in a more culturally relevant manner."

—Donald Easton-Brooks, Associate Dean of the School of Education, Hamline University

"This book is a wake-up call for faculty to critically examine the extent to which corporations, conservatives, quasi-mental health professionals, and the military have infiltrated higher education in a quest to suppress a liberal education—our raison d'etre. Students and faculty are scrutinized for their behavior, and emergency response behavioral teams are kicked into action at the slightest indication of a mental health issue. 'Deviance' is back in fashion, and we are all at risk from this new SWAT team."

—Janet Duncan, Associate Professor, Foundations and Social Advocacy,
School of Education, SUNY Cortland

"A must-read for professors, students, staff, administrators, and the general public concerned about the future of education and democracy. Finally, here is a book that examines the police state that is growing on every campus."

—Priya Parmar, author of *Knowledge Reigns Supreme: The Critical Pedagogy of Hip-Hop Artist KRS-ONE*

"*Policing the Campus* is a pointed collection that takes aim at the disciplinary logics and practices that increasingly dominate higher education. The essays that explore the relationships between campus activism and the Occupy Movement are especially timely, but the entire book is a fruitful contribution to the debate over the freedom of the university in the twenty-first century."

— Zack Furness, editor of *Punkademics*

POLICING THE CAMPUS

ACADEMIC REPRESSION, SURVEILLANCE, AND THE OCCUPY MOVEMENT

EDITED BY ANTHONY J. NOCELLA II
AND DAVID GABBARD

PETER LANG
New York • Washington, D.C./Baltimore • Bern
Frankfurt • Berlin • Brussels • Vienna • Oxford

Library of Congress Cataloging-in-Publication Data

Policing the campus: academic repression, surveillance, and the Occupy Movement /
[edited by] Anthony J. Nocella II, David Gabbard.
pages cm. — (Counterpoints: studies in the postmodern theory of education, vol. 410)
Includes bibliographical references and index.
1. Campus violence—United States.
2. Universities and colleges—Security measures—United States.
3. College students—United States—Political activity.
4. Campus police—United States. 5. Electronic surveillance—United States.
6. Occupy movement—United States.
I. Nocella, Anthony J., editor of compilation. II. Gabbard, David.
LB2345.P65 378.1'961—dc23 2013003346
ISBN 978-1-4331-1312-3 (hardcover)
ISBN 978-1-4331-1311-6 (paperback)
ISBN 978-1-4539-1038-2 (e-book)
ISSN 1058-1634

Bibliographic information published by **Die Deutsche Nationalbibliothek**.
Die Deutsche Nationalbibliothek lists this publication in the "Deutsche
Nationalbibliografie"; detailed bibliographic data is available
on the Internet at http://dnb.d-nb.de/.

Cover image of John Pike using pepper spray used with permission by Louise Macabitas.

The paper in this book meets the guidelines for permanence and durability
of the Committee on Production Guidelines for Book Longevity
of the Council of Library Resources.

29 Broadway, 18th floor, New York, NY 10006
www.peterlang.com

Printed in the United States of America

This book is dedicated to those who have been silenced, marginalized, repressed, not accepted, can't afford, or are not allowed in college or university. This book is for all of those who demand a free democratic critical inclusive education for social justice, void of guns, police, security, and surveillance.

Contents

PART III
FROM DEFENDING PUBLIC EDUCATION TO THE OCCUPY MOVEMENT

Acknowledgments

Acknowledgments are so important because it is impossible to be here or do anything on this planet without the support of others. Anthony Nocella III would like to thank first and foremost Connie, Justin, and everyone at Peter Lang especially Chris Myers, Bernadette Shade, and Stephen Mazur. We would also like to thank Christian Parenti for writing the wonderful foreword, and all the contributors to the book: Natalia Abrams, Sarah Augusto, Ben Brucato, Jason Del Gandio, Scott DeMuth, Luis Fernandez, Henry Giroux, Daniel White Hodge, Caroline Kaltefleiter, Joe Lewis, Barbara Madeloni, David Pellow, Julie Setele, Kimberly Socha, Maura Stephens, Wesley Strong, Ryan Thomson, and Rich Van Heertum; for if they were not involved, this book would not be possible. We would also like to thank, in no particular order, Zack Furness, sj Miller, Priya Parmar, Sarat Colling, Susan Thomas, Renee L. Wonser, Donald Easton-Brooks, Steven W. Halady, John Alessio, Janet Duncan, Julie Andrzejewski, and Bill Templer, who have supported this book by reviewing it. We would finally like to thank our families and our many friends, colleagues, universities, centers, institutes, and organizations.

David Gabbard would like to thank his family for all of the love, support, and understanding over the years. He would also like to thank Anthony Nocella II for his patience and understanding, and for bringing him on board this project to begin with.

Anthony J. Nocella II would first like to thank his family members who always say they come last in every acknowledgment. Next he would like to thank David Gabbard for being an amazing friend and co-editor of this great important critical book. He would also like to thank his advisors, Peter Castro, Richard Loder, Micere Mugo, Tucker Culbertson, Piers Beirne, Kishi Ducre, Barbara Applebaum, and Dalia Rodriguez at Syracuse University. Anthony would also like to thank all of his friends and colleagues at SUNY Cortland and Le Moyne College for their support and care. He would like to thank everyone with Save the Kids in and outside the walls of detention facilities;

everyone with the Institute for Critical Animal Studies, especially the boardmembers, Susan Thomas, Carolyn Drew, Sarat Colling, Amber Gilewski, Stephanie Jenkins, Helena Pedersen, Les Mitchell, Colin Salter, Vasile Stanescu, Nicola Taylor, Richard Twine, Richard White, Felipe Andrusco, Kimberly Socha, and Dylan Powell; everyone with the Central New York Peace Studies Consortium; the *Peace Studies Journal*, the *Journal for Critical Animal Studies*; the *Journal for Critical Urban Education*; the Center for Excellence in Urban Teaching; the Institute for Hip Hop Knowledge; Hillbrook Youth Detention Facility; Hamline University, and of course, the Hamline School of Education. Finally my friends in no order: Ernesto Aguilar, Mecke Nagel, Donald Easton-Brooks, Matt Hernandez, Lara Drew, Laura Shields, Jay Grims, Alma Williams, Kevin Beacham, Naomi Taylor, Jean Strait, Rachel Endo, Tammi Pate, Yvonne Wells-Ellis, Rob Rivera, Frank Hernandez, Tessa Tiogress, Moses Jame, Ankit J. Sheth, Stephanie Robinson, Taylor Andrew, Peter McLaren, Brian Trautman, Liat Ben-Moshe, Don Sawyer, James Czarniak, Jackie Denero, Catherine Garrone, and Hasan Stephens. Anthony would like to especially remember and thank his late best friend and roommate, who was loved by many, Kevin Pieluszczak.

• FOREWORD •

This Is Your Mind on Lockdown

Christian Parenti

In 2008, campus police at the Old Westbury Campus of the State University of New York on Long Island put surveillance cameras in the hallways of the Woodlaw dormitories. Costing approximately $700,000, the 11 high-tech German-made cameras have powerful zoom-capability and can record sound, though campus police claim they do not use that feature. Now young women transiting from their rooms to the showers at the end of the hall, often naked except for towels wrapped around them, do so under ever-present police gaze.

The students at this predominantly African American campus have complained about the cameras, saying they feel violated and slightly threatened, and have consulted the ACLU. But apparently this type of legally questionable surveillance is too hard to fight. In other words, these students in effect have no Fourth Amendment expectation of privacy at the threshold of their most intimate spaces, their bedrooms and bathrooms. And so, they grow accustomed to life under hostile observation.

Along with the Old Westbury dorm cameras come all the shaming and isolation that excessive policing confers on its target populations. "We feel there are certain things we can't talk about," said one dorm resident to student journalist Selena Hill. "I am being watched and I don't want to feel that way." In that regard, the cameras are just the latest hallmarks of old patterns of social control and regimentation in a class society structured in part along the lines of racialized hierarchy that has long kept the poor and working classes, and particularly people of color, in check by means of surveillance, threats of violence, and actual force both in workplaces and communities.

Old Westbury's dorm cameras are an inconvenience and humiliation for many of those living under them, but the threat goes far beyond the personal sensibilities of some students; the cameras (or rather the politics they embody) threaten society as a whole. In fact, the political threat these cameras pose might be better measured by the number of students who *do not* feel uncomfortable being watched. There is a firm link between the life of the mind and the politics of the body. One does not need to read Antonio Gramsci or Michel Foucault (though it helps) to understand that if your mind is shackled, so too is your body. And conversely if you surrender control of your body—allowing it to be measured, searched, observed, categorized, pushed hither and thither by police and administrative authorities—so too will your mind eventually lose its autonomy.

Routine surveillance is like water torture upon the will: constant, slow, gentle but forced submission to authority; surveillance habituates us to obedience. A population with no sense of individual and collective autonomy is a population incapable of political action and physical mobilization.

Because of the university's unique role in political life, campus policing comes with broad political implications. Universities in the West have a long history of autonomy—financial, political, and academic. Without politically and intellectually autonomous universities, society loses a crucial wellspring of criticism and innovation; and a vanguard terrain of political education, action, and mass mobilization. Modern American universities have been pivotal places from which movements and radicalized progressive communities have been born. Over-policing threatens that. The essays in this excellent volume explore these issues in harrowing detail.

Lest anyone has forgotten, progress as we know it is largely the result of protests. Consider the list of achievements that we think of as progress: the slow and steady emancipation of women and people of color from social and political controls, the end of child labor, the rising standard of living for working-class people from the beginning of the 20th century until the mid-1970s; the advent of effective health, safety, and environmental regulations—all these are at least partially the product of protest. In the current United States of America—stalked by right-wing Christian fanaticism, rising economic inequality, and a profound multifaceted environmental crisis—if there is not broad political mobilization demanding social and environmental justice, we will face a very bleak future.

It will be hard for a generation raised under constant surveillance and police bullying—in the form of metal detectors, bag searches, stop and frisk—to arrive at university only to receive more of the same, to then develop the mental and cultural autonomy necessary to build effective protest movements and achieve progressive social change.

What is at stake in policing the campus is more than just the campus. This book addresses some of the most vital issues of our time: as individuals, can we live free and dignified lives? As groups, will we be free to think, debate, dissent, assemble, organize, and challenge powerful interests with protest?

Canary in the Coal Mine?

David Gabbard

February 20, 2005. Hunter S. Thompson sits alone in his favorite chair in the kitchen of his home at Owl Farm in Woody Creek, Colorado. His son Juan and his family are visiting from out of town, but they're off in another room. Anita, his wife, is at the fitness club. Hunter had always lived by the motto that "It's wrong when it stops being fun." For Hunter, fun had always been the object of life, including work. For those of us who came of age reading the products of his brilliance, fun was always part of the equation. Of course, Hunter was a political journalist and a person of tremendous personal integrity, so telling the truth was always paramount. He was going to tell the truth, no matter how much fear and loathing it might inspire. But maybe it was because of the fear and loathing that he didn't want to tell it to us straight, and why he had to cut it with something. For Hunter, that something was *fun*. And this mixture of hardcore political truth and fun constitutes the essence of what fellow journalist Bill Cardoso dubbed Hunter's "Gonzo" style of journalism. Sadly, maybe even tragically, too many of Hunter's readers prize the truths of Hunter's insights less than they do the fun he brings to their telling. Readers in the future ought to remember that Hunter didn't embrace Gonzo merely for the sake of being outrageous, but because the term reminded him of William Faulkner's observation that the best fiction is far truer than journalism (Steadman, 2006, 69).

But the fun had started fading from Hunter's life. Over the past few years, he'd undergone a wrenching period of detox from alcohol addiction, a hip replacment, a broken ankle, and two back surgeries. In addition to his notorious pattern of hard and fast living, these highly invasive procedures took a heavy toll on his health, as well as his spirit. Hunter was losing what meant most to him—his independence—and he could feel himself becoming an increasing burden to his family. He *hated* it; and he was having none of it.

He had broached the topic with Anita before, and she'd grown furious with him, which she later regretted. In retrospect, she knew that he couldn't go on like this. His dignity wouldn't allow for it. It was just *wrong*.

Four days earlier, he crafted the note, titling it "Football Season Is Over":

> No More Games. No More Bombs. No More Walking. No More Fun. No More Swimming. 67. That is 17 years past 50. 17 more than I needed or wanted. Boring. I am always bitchy. No Fun—for anybody. 67. You are getting Greedy. Act your old age. Relax—This won't hurt. (2005)

The time has come. He calls Anita at the fitness club and asks her to come home to help him work on a writing assignment from ESPN. With her still on the line, he sets the phone down on the table. She hears the metallic clicking as he chambers the round. He puts the barrel of the .45 automatic into his mouth and squeezes the trigger. Just like that, the thing is done. Hunter S. Thompson is dead.

We have to avoid the temptation to play forensic psychologists. "Was it really his lost sense of independence resulting from his declining health that led Hunter to end his own life? Or was there something else?" We can't really know for certain. However, at the end of his book chronicling his 35-year professional and personal relationship with Hunter, Ralph Steadman (2006) includes a posthumous open letter to his friend that helps explain the title of our introduction. Steadman writes,

> It is over a year since you set the scene for your strangely planned suicide. The messages I had received from you up to the Bush-Kerry election of 2004 were infected by a dismal undertone. "If Bush wins," you said, "the planet is doomed. The Halliburton Corporation will rule with an army of obedient geeks who will transform America into a land fit for religious freaks, informers, fascist legislators, greedheads, an obscene defense budget against the "terrorists" and the desire for pre-emptive action against Iran and the ambition to remain the most dominant power on earth by whatever means." (2006, 383)

The letter turns on Steadman's account of Hunter's request that Steadman, who had illustrated all of his books since *Fear and Loathing in Las Vegas* (1971), draw a picture that would capture "the personification of absolute evil" (Ibid.), and his inability to effectively honor that request while they were together at the ranch of a mutal friend in October of 2004 before the election. "It was a conundrum," Steadman explained, "to be solved after you were gone"—that is, *after* Hunter's death.

> Why shouldn't I think that this was one of those times when I could deliver like before, do the proper thing? Kick ass or get kicked in the ass? Well, we got our asses kicked. The world got its ass kicked. Although I tried to give my best on this occasion, it wasn't enough. Cartoon satire was not a cure-all answer to anything, particularly regarding the election of the President of the United States.

Finally grasping what was (at least for Hunter) the personification of absolute evil, Steadman states that

> The President embodies all that is wrong with our planet, our environment and our predicament, and your beloved nation was duped into voting for him by corporate deceit. Bush is the first President to do this (save for our old friend Richard Nixon whom you had begun to consider as a barrel of laughs compared to this one).
>
> George W. Bush is a fraud who beat Al Gore by cheating and won a second term by presenting his credentials in a violent lie as an enemy of terrorists and the saviour of democracy throughout the world. He is scum and always will be scum. He will live on as the lowest form of intelligence at an executive level this world has had to suffer.
>
> Hunter, that is why you committed suicide. Your America was gone. It was seriously the death of fun. THE JOKE WAS OVER. (Ibid., 384)

In making his decision, Hunter never meant to martyr himself to any democratic idealism. There was nothing in his note to suggest that it was anything other than a very private decision—one he'd given a great deal of frequent thought to over many years. He once told Steadman that "I'd feel trapped in this life if I didn't know I could commit suicide at any time" (Steadman, 2006, 2). Therefore, it's only in retrospect that we can judge him as being like a proverbial canary in a coal mine. The spirit that once animated him was a casualty of his own personal war against injustice, calumny, greed, sloppiness, stupidity, and brutality. Like Steadman said, "we got our asses kicked," and those of us who continue that fight are still having our asses kicked. For its part, though it was never intended to, this book chronicles some of the latest episodes in that ongoing ass-whupping, bearing witness to much of what Hunter prophesized in the final years of his life.

In his author's note to *Kingdom of Fear*, Hunter wrote

> I like this book, and I especially like its title, which pretty well sums up the foul nature of life in the U.S.A. in the first few bloody years of the post-Amerian century. Only a fool or a whore would a call it anything else.
>
> It would be easy to say that we owe it all to the Bush family from Texas, but that would be too simplistic. They are only errand boys for the vengeful, bloodthirsty cartel of raving Jesus-freaks and super-rich money mongers who have ruled this country fror the last 20 years, and arguably for the past 200. They take orders well, and they don't ask too many questions.
>
> The real power in America is held by a fast-emerging new Oligarchy of pimps and preachers who see no need for Democracy or fairness or even trees, except for maybe the ones in their own yards, and they don't mind admitting it. They worship money and power and death. Their ideal solution to all the nation's problems would be another 100 Year War.

> *Coming of age in a fascist police state will not be a barrel of fun for anybody, much less for people like me, who are not inclined to suffer Nazis gladly and feel only contempt for the cowardly flag-suckers who would gladly give up their outdated freedom to live for the mess of pottage they have been conned into believing will be freedom from fear.* (emphasis added)
>
> Ho ho ho. Let's not get carried away here. Freedom was yesterday in this country. Its value has been discounted. The only freedom we truly crave today is freedom from Dumbness. Nothing else matters. (2003, xxi–xxii)

He might have viewed the election of Barack Obama to the presidency as a step in the right direction—away from the Dumb. On the other hand, his political connections and intuitions could just as well have cautioned him against any high expectations. He wasn't naïve about the machine that runs Chicago politics, and he would have been amused to see Rahm Emanuel leave his position as Obama's Chief of Staff to assume control of that machine as the city's mayor. In that case, he would not have been terribly surprised by the brutality of the legal and police tactics taken against the Occupy Wall Street uprising. After all, Hunter attended the 1968 Democratic Convention in Chicago, where he was gassed, arrested, and had his nose broken by Mayor Daly's police. And he had an intuitive sense that, as bad as things had gotten after just the first four years of George W. Bush, things were only going to get worse. Just 18 months prior to his death, he wrote:

> The American nation is in the worst condition I can remember in my lifetime, and our prospects for the immediate future are even worse. I am surprised and embarrassed to be a part of the first generation to leave the country in far worse shape than when we first came into it. Our highway system is crumbling, our police are dishonest, our children are poor, our vaunted Social Security, once the envy of the world, has been looted and neglected and destroyed by the same gang of ignorant, greed-crazed bastards who brought us Vietnam, Afghanistan, the disastrous Gaza strip, and ignominious defeat all over the world.
>
> The Stock Market will never come back, our Armies will never again be Number One, and our children will drink filthy water for the rest of their lives.
>
> The Bush family must be very proud of themselves today, but I am not. Big Darkness, soon come. Take my word for it. (2005, 222)

"Are We There, Yet?"

No, we have not hit the bottom. Though few would admit it—or even recognize it, we're still in free fall, plummeting deeper and deeper into the Big Darkness of our abyss. As suggested earlier, this book chronicles scenes from that descent as it is currently being played out on college campuses across our country. In conceptualizing the volume, we sought contributions that would address what we view as a general criminalization of dissent and activism. The chapters in Part I provide a broad overview of those issues, beginning with Jason Del Gandio's essay that situates current trends within the framework of neoliberalism and the corporatization of higher education. In his analysis, neoliberalism represses the formation of any civic imperative that might drive the mission of our institutions of higher education to serve the public welfare of common good. Within the neoliberal framework, there is no public space and no common interest. There are only the private interests of the competitive individual, acting and struggling alone within the economic sphere. Colleges and universities must, therefore, restrict their focus to specialized workforce training. Wesley Strong extends this line of argumentation in Chapter 2, describing four methods by which corporatization represses the civic imperative in higher education: policy, nonviolent police action, police violence, and legal action. In addition to his analysis of how those forces impact the civic development of students, he also describes how neoliberal policies exploit students and contribute to their growing debt burden.

Daniel White Hodge examines in Chapter 3 how minority students, particularly African American and Hispanic males, experience these repressive forces on college campuses much more intensely than others. In the eyes of too many campus police, the mere color of one's skin signifies a threat to campus safety. Consequently, as a microcosm of the dynamics of race in the larger society, life on campus for minority students feels like life in a police state. In Chapter 4, Kim Socha brings faculty into the conversation through her examination of *policed pedagogy*. As she explains, policed pedagogy refers to teaching under external pressures that can stem from three different sources: administration, law enforcement, or organized propaganda campaigns aimed at intimidating faculty into curricular conformity with some larger political or cultural agenda.

Illustrative of Socha's points, in Chapter 5, Barbara Madeloni chronicles her own experiences at the University of Massachusetts, where the administration recently refused to renew her contract in retribution for her role in challenging the university's requirement for students in pre-service teacher education to participate in a nationwide field study run by Pearson Inc. of the Teacher Performance Assessment developed at Stanford University.

The chapters comprising Part II look at how college campuses are coming to mirror the signature features of the larger "surveillance society." Richard Van Heertum, in Chapter 6, describes the use of surveillance cameras and ID card swipes. In his analysis, Michel Foucault's concept of the panopticon maintains its relevance to the situation, but he also considers the deployment of post-panoptic and synoptic techniques of surveillance. In Chapter 7, Ben

Brucato and Luis A. Fernandez pick up the Foucauldian analysis of contemporary campus policing strategies, connecting them to broader social and discursive histories that help contextualize particular socio-technical innovations that link surveillance, access-control, and weaponry. Caroline K. Kaltefleiter in Chapter 8 also relies heavily on Foucault's work in her discussion of the development of campus alert systems and federal legislation intensifying government surveillance of citizens.

David Pellow and Scott DeMuth pick up this Foucauldian theme of surveilllance in Chapter 9, linking it to the state's use of repression as a rational science that begins with a demand to know. More specifically, Pellow and DeMuth explore the manner in which the federal Animal Enterprise Terrorism Act (AETA) functioned to facilitate a case of academic repression against DeMuth and other scholar-activists aligned in solidarity with the animal liberation movement. In Chapter 10, Joe Lewis expands the scope of this area of inquiry, exploring the broader theme of how new security technologies and the culture of surveillance are undermining free inquiry on campuses across the country.

The chapters in Part III close out the book with a focus on how university administrations and other authorities have responded to the presence of the Occupy Movement on campus. Sarah Augusto and Julie Setele provide an in-depth account of events leading up to and following the pepper-spray incident at the University of California–Davis, tying that incident and the policing of today's campus protests to the privatization of higher education and the militarization of (campus) police. In Chapter 12, Maura Stephens helps readers recognize that such incidents are far from isolated. In her view, they form a pattern of a larger and far more dangerous pattern of repression aimed at silencing dissent on campuses, rendering them safe spaces for the further corporatization of education.

As if forged in direct response to Stephens's plea for members of university communities to become more engaged in disrupting the patterns described in hers and previous chapters, the final two chapters of the book either answer or echo this call for resistance. In Chapter 13, Ryan Thomson and Natalia Abrams provide an insider's view of the courageous work of the Occupy Colleges Movement. A student-led movement, Occupy Colleges strives to challenge the corporatization of higher education on a number of fronts. In Chapter 14, noted educational and social theorist Henry Giroux closes the book with a passionate call for more university faculty to join them.

Conclusion

In Fear and Loathing in Las Vegas (1972), Hunter wrote:

> History is hard to know, because of all the hired bullshit, but even without being sure of "history" it seems entirely reasonable to think that every now and then the energy of a whole generation comes to a head in a long fine flash, for reasons that nobody really understands at the time—and which never explain, in retrospect, what actually happened. (67)

For Hunter and the members of his generation, that energy came to a head in the middle of the 1960s, a period that is still hated by our oligarchs and their courtiers. Reflecting on that moment, he wrote:

> It seems like a lifetime, or at least a Main Era—the kind of peak that never comes again. San Francisco in the middle sixties was a very special time and place to be a part of. Maybe it meant something. Maybe not, in the long run . . . but no explanation, no mix of words or music or memories can touch that sense of knowing that you were here and alive in that corner of time and the world. Whatever it meant . . . .
>
> You could strike sparks anywhere. There was a fantastic universal sense that whatever we were doing was right, that we were winning. And that, I think, was the handle—that sense of inevitable victory over the forces of Old and Evil. Not in any mean or military sense; we didn't need that. Our energy would simply prevail. There was no point in fighting—on our side or theirs. We had all the momentum; we were riding the crest of a high and beautiful wave.
>
> So now, less than five years later, you can go up on a steep hill in Las Vegas and look West, and with the right kind of eyes you can almost *see* the high-water mark—that place where the wave finally broke and rolled back. (1972, 68)

The contributors to *Policing the Campus* make it clear that "the forces of Old and Evil" are in no mood to tolerate dissent on campus any more than they tolerate it elsewhere. For them, the Occupy Movement poses a much greater threat to the hegemony of corporate oligarchy than the campus movements of the 1960s ever did. Those movements had little to no support beyond the universities. With Occupy, it's the campus that's joining a movement that started on the streets, Wall Street to be exact. If the working people associated with the Tea Party ever forge an alliance with Occupy and the numerous student branches of Occupy, that hegemony could be in real danger. And so the clampdown has come down early and often. "Old and Evil" is much better prepared, and far more heavily armed, this time around.

It's sad that Hunter didn't live to see the Occupy Movement. If history is hard to know, then the future is even harder. We still don't know what it all means. Our energy hasn't come to a head yet, but any hope of moving the American democratic experiment forward will require all of us to keep pushing to build the momentum started on September 17, 2011, to ride our own "high and beautiful wave."

Bibliography

Steadman, R. (2006). *The Joke's over*. New York: Harcourt.

Thompson, H. S. (1972). *Fear and loathing in Las Vegas*. New York: Simon and Schuster.

——. (2003). *Kingdom of fear: Loathsome secrets of a star-crossed child in the final days of the American century*. New York: Simon and Schuster.

——. (2005). *Hey rude: Blood sport, the Bush doctrine, and the downward spiral of dumbness*. New York: Simon and Schuster.

PART I

Campus Police

• CHAPTER ONE •

Arrests and Repression as a Logic of Neoliberalism

Jason Del Gandio

It's a beautiful, sunny weekday afternoon on college campus. People are mingling about, moving to and from classes, parking lots, the student center, the library, the cafeteria, and administrative offices. Lectures and class discussions are heard through the windows; students and teachers chat beneath the trees; fraternities and sororities seek new recruits; and extracurricular programs promote community service projects. Despite this charming scene, the wider world rages on: the Gulf Coast oil spill destroys communities, wildlife, and the natural environment; climate change continues with little conscience from the G7; people of color are racially profiled under the guise of "illegal immigration"; same-sex couples are denied marriage and the conjoining rights of tax breaks, hospital visitation, Social Security, and spousal retirement benefits; lower- and working-class citizens, as well as students, nonprofits, social services, and art and cultural programs are disproportionally affected by the financial crisis; swaths of people face the possibility of homelessness as unemployment rises and homes foreclose; and the U.S. military continues to receive $700 billion annually.

After much thought and discussion, groups of students, faculty, and staff determine that their college campus does not exist within a vacuum–it is neither an idyllic island nor a neutral site of intellectual reflection. Instead, it is a microcosm that reflects and is reflected by society. The college campus thus provides an opportunity–even a moral obligation–to combat wider ills and champion wider goods.

These campus community members decide to transform their college into a site of political contestation. They organize a protest around a slogan adopted from the 2010 United States Social Forum: "Another world is possible! Another U.S. is necessary! Another campus is happening!" A few hundred people attend, with many more observing and inquiring. Every imaginable issue is addressed–war, public education, immigrant rights, racism, sexism, tuition, campus politics, local elections, political imprisonment, the environment, etc. Rather than signaling disorganization, confusion, or mixed-messaging, this multi-pronged approach rests upon two political concepts: intersectionality, which is the idea that numerous issues overlap into overarching oppressions; and *specifismo*, which is the Latin American idea that different political positions can stand in tenuous and contentious yet healthy and productive solidarity.

Preparing for the event serves as a course in hands-on political praxis—and not just for the students, but also for the participating faculty, staff, and wider community members. People learn how to coordinate and publicize a large event; how to negotiate different personality types; how to set and achieve short- and long-term goals; how to communicate one on one, in small groups, and via technology; how to navigate bureaucratic processes; how to adjust to changing demands; and most of all, how to fight for a better world.

The organizers have worked for months and the day finally arrives. People march loudly and joyously through campus on their way to a designated area. A vibe fills the air and a vision fills their hearts. The collective experience intimates an alternative form of power—not power *over* others, but power *with* one another and power *against* the institutions, structures, and discourses that enclose, disable, and paralyze this very kind of action (Starhawk, 1987). As political theorists Michael Hardt and Antonio Negri might say, the protesters embody an open and creative *constituent power* that antagonizes and undermines the capture and control of *constituted power* (2010, 22–38). It's people power versus state power; creative energy versus static energy.

But not everyone is satisfied with this form of protest. Police escorts, free-speech zones, and orderly demonstrations do not necessarily threaten the ills of society. A smaller group of students thus organize an alternative action—the occupation of the president's office. At a predesignated time, a dozen people storm into the building. All staff are asked to exit. The students then lock and barricade the doors, cut the phone, Internet, and security camera lines, and drop a banner from the roof that reads, "A Free Education for a Free People, NOW!" Referring to themselves as the "Autonomous Student Insurrection" (ASI), they communicate their demands via cell phones and portable wireless Internet. Their action appropriates physical space, public messaging, the use of communication technology, and of course, political agency. Their refusal to negotiate or compromise is an enactment of their temporary autonomy. In their eyes, college campuses and the pursuit of education will never be free as long as administrations—backed by the force of the law—are in a position to grant the right to protest. In a reversal of perspective, ASI argues that the granting and revocation of privileges is inherently oppressive. ASI does not seek the abolition of all rules, but instead, the abolition of top-down rulings that serve the interests of power elites.

School authorities quickly contact the local police and make decisions. The building is eventually surrounded by riot police, yellow tape, metal barricades, police vans, fire trucks, ambulances, paramedics, and mass media. The scene is more reminiscent of a bank robbery, natural disaster, or virus outbreak than a college campus. With ASI refusing to exit, the police force their way into the building. A few of the students attempt to make a break for it, but are pepper sprayed and wrestled to the ground. The remaining students crowd together and lock arms. Although they maintain their commitment to nonviolence, they repeatedly chant "Free Education, Free People! . . . Free Education, Free People!" The police strip ASI apart one by one, placing each student into plastic handcuffs and marching them out to carrying vans. The official charges: trespassing, property damage, and illegal occupation. The possible punishments: fines, community service, expulsion, and jail time.

This scenario may be hypothetical, but it's not far from reality. Protests, direct actions, arrests, and police clashes regularly occur across college campuses. Two of the more recent and dramatic examples demonstrate the point.

Police Repression

In December of 2008, a group of students occupied a cafeteria at the New School in New York City. They issued a host of demands including but not limited to the resignation of university president Bob Kerrey, executive vice president James Murtha, and treasurer of the board of trustees Robert Millard. The students argued, among other things, that these university administrators did not reflect the progressive traditions of the New School and strategically inhibited wider student participation within university operations. The occupation lasted approximately 30 hours without arrest. Although several negotiations were made, a student group referring to itself as New School in Exile warned of further actions if President Kerrey did not resign by April 1 (Moynihan and Chan, 2009; New School in Exile, n.d.; Wharton, 2008).

New School in Exile followed through with its warning. On April 9, 2009, a group of students occupied another building. This time, school administrators sent in the police, which led to a scenario very similar to the one described above. Twenty-two people were arrested, a few pepper sprayed, and one street videographer was assaulted by a police officer. When asked about the situation, President Kerrey stated (among other things), "This is not a university situation, this is an NYPD situation. The students broke and entered into the building.... I cannot consider them students at this point" (Del Signore, 2009; Fleming, Doyle, and Nichols, 2009).

On the other side of the country, local and statewide protests occurred in California from September, 2009 through May, 2010. The 2008 financial crisis, combined with the state's ongoing financial problems, led to crippling budget cuts for higher education. The community college, Cal State, and U Cal systems began implementing fee increases (equivalent to tuition), pay cuts, layoffs, enrollment waiting lists, and mandatory temporary leaves of absence for all university employees. The state's higher-education system, which was close to being free at one time, is now being gutted. Major protest actions occurred at Berkeley, Santa Cruz, San Francisco, Fullerton, Sacramento, Davis, and Los Angeles. Actions included but were not limited to sit-ins, teach-ins, and study-ins; walk-outs, hunger strikes, and tent cities; marches, rallies, and dance parties; building occupations, traffic disruptions, and campus blockades; property damage and fires. Numerous arrests ensued. Actions at Berkeley were particularly intense when a protest moved off campus to downtown. A crowd of more than 200 eventually clashed with police, which led to rioting, broken windows at corporate chains, and burned dumpsters and trash cans. One of the commonly used slogans spotlights the tension: "California #1 in Prison Spending, #48 in Education" (Johnson and de Vise, 2010; McKinely, 2010; Occupy California, n.d.; O'Leary, 2009; Wollan and Lewin, 2009).

The New School and California protests are but two of the more extreme and well-publicized examples of student protest and police repression. In August 2001, Miguel Malo, vice-president of student government at Hostos Community College, was violently arrested after single-handedly protesting cuts to bilingual programs and registration fee increases. In September 2005, George Mason University student and Air Force veteran Tariq Khan was assaulted by police and then charged with trespassing and disorderly conduct after staging a peaceful, one-person protest against military recruitment on campus. In September 2007, University of Florida student Andrew Meyer was tasered and arrested after repeatedly asking presidential candidate John Kerry controversial questions. In April 2010, the University of Washington police department authorized an undercover police officer to attend, participate in, and collect information about a meeting of the university's Student Worker Coalition.

The list could go on and on, but it is obvious that student protests are commonly met with police repression. More times than not, school administrations authorize such repression. It is not my intent to demonize or essentialize administrators. Administrators play an important role in facilitating college life and each individual fulfills that role through different values, beliefs, and practices. However, administrators—as well as students, faculty, and staff—exist within wider socio-political forces that establish the conditions for the possibility of particular actions and decisions. This chapter unravels some of the forces that motivate repression of student protest.

In many ways, school administrations have no choice but to preemptively inhibit and if need be forcibly stop such political disruptions as lock-ins, occupations, blockades, and unpermitted marches. Administrators are responsible for maintaining school operations and the well-being of *all* community members. This standard response seems reasonable. But it's reasonable only within a logic that situates student protesting as a threat to campus life. Who and/or what is actually threatened and/or protected? I argue that repression of student protesting is supported and even demanded by the socio-political logic of neoliberalism. This is not to suggest that such repression is new or unique to the 21st century. The campus repression of the 1960s/70s was probably more intense—or at least more outwardly violent—than that of today. But different eras engender different socio-political logics that rationalize and mobilize such repression. Neoliberalism is the logic of today.

Neoliberalism and Corporate U.

Plenty of literature supports the argument that colleges are increasingly corporatized (Aronowitz, 2000; Giroux, 2002; Giroux and Myrsiades, 2001; Edu-factory, 2009). According to Jennifer Washburn, author of *University Inc.* (2005):

> In higher education today, a wholesale culture shift is transforming everything from the way universities educate their students to the language they use to define what they do. Academic administrators increasingly refer to their students as consumers and to education and research as products. They talk about branding and marketing and now spend more on lobbying in Washington than defense contractors do. Many have eagerly sought to convert "courseware" into intellectual property that can be packaged and sold over the Internet for profit. Others have allowed whole academic departments to forge financial partnerships with private corporations, guaranteeing these firms first dibs on the inventions flowing out of their labs. (ix–x)

Such corporatization transforms higher education from a facilitator of worldly experience to job preparation for the new global economy. According to a report by the National Education Association (2004), colleges increasingly cater to vocationally oriented programs at the expense of the liberal arts. Since 1970, the number of degrees in English and literature declined by 13,000 and the number of degrees in foreign languages by 6,000. Meanwhile degrees in business, accounting, computer science, and mass communications flourish. Even former Harvard president Derek Bok (2003) acknowledges that "the surrounding economy draws more and more students into vocational fields of study. . . and attracts greater sums of money for subjects of commercial relevance to the neglect of other worthy, but less practical, fields of study" (17).

This corporatization is driven by neoliberalism. In brief, neoliberalism is a form of global capitalism based on the deregulation of free markets and the privatization of wealth. It subordinates government control to the interests of private wealth accumulation. The government—rather than regulating the market to ensure fair competition—becomes an extension of market activity, a servant of the industries to which it is captive. Neoliberalism provides tax breaks for the rich, reduces spending on social programs and welfare, expands corporate control, and eradicates labor rights, environmental protections, drug and food regulations, and even national law.

This "neo laissez faire" approach allows private interests to own and control every aspect of the human, social, and natural world. Things like food, water, farmland, forests, healthcare, prisons, militaries, political processes, mass media, and, of course, education, are targets of neoliberalism. Even individual thoughts, plant seeds, mothers' breast milk, and human DNA are intended to be owned, controlled, bought and sold by free-market capitalists.

While most colleges are still nonprofit institutions, their primary function is to serve the economic aspirations of neoliberalism. This happens on at least three levels. First, campus communities are transformed into market populations through licensing agreements; construction contracts; outsourcing of bookstores, vendors, concessions, and food, laundry, traveling, and printing services; corporate sponsoring of buildings, events, speakers, and campus programs; patenting of intellectual property rights; and funding, ownership, and direct influencing of research (Wertz, 2000; Rae-Dupree, 2008).

Second, the overall educational enterprise is funneled into, and functions synonymously with, privatized wealth accumulation. For instance, the percentage of high school graduates attending college rose from 42 percent in 1970 to 70 percent in 2009 (Bureau of Labor Statistics, 2010). The socio-economic worth of a college degree also increased during this time period. In 1980, the weekly salary of college graduates was 40 percent higher than that of high school graduates. By 1997, that gap had risen to 73 percent (Levine, 2001). These statistics could be interpreted as a progressive shift toward a more educated and prosperous society. But economic inequality actually increased over these years. In 1979, the top 1 percent of Americans owned 20.5 percent of the nation's wealth, while the bottom 99 percent owned 79.5 percent. By 2007, the top 1 percent increased its share to 34.6 percent while the bottom 99 percent declined to 65.4 percent (Domhoff, 2010). In 1980, the pay ratio between the average American CEO and average American worker was 40 to 1. As of 2007, the ratio was 364 to 1 (Anderson et al., 2007). These additional statistics demonstrate that higher education helps the individual move upward *at the expense* of other individuals—i.e., academia contributes to both upward mobility *and* wider social stratification. The smarter we get, the more unequal we become. Such private individual gain, rather than public collective gain is part and parcel of neoliberalism.

This Social Darwinism segues into the third and perhaps most important point: neoliberalism acts as a public pedagogy that teaches people how to think, act, believe, and live. According to educational theorist Henry Giroux, "Neoliberalism not only transformed economic agendas throughout the over-developed world; it also transformed politics, restructured social relations, and produced an array of reality narratives (not unlike reality TV) and disciplinary measures that normalized its perverted view of citizenship, the state, and the supremacy of market relations" (2010, 101). Neoliberalism is not simply an economic system, but a socio-political discourse that thoroughly permeates the mental and behavioral landscape. Tacit presuppositions about the relationship between free-market capitalism and social well-being are taken as immutable facts. People at all levels of society assume, for instance, that government is subordinate to capital, that democracy is reducible to consumer choice, that technological innovation can solve social and ecological crises, that individual identity is akin to corporate branding, and that rich people earn it while poor people deserve it. More troubling than these explicit statements are the implicit values: competition is better than cooperation, egotism better than gratitude, personal prestige better than community service, control better than assistance, self-help better than solidarity, and fortune better than spirit.

Giroux argues that this "neoliberal common sense" coexists with a logic of punishment (2009; 2010). Many Americans believe, for instance, that the 2008 mortgage crisis was a result of financially irresponsible homeowners rather than corrupt and unregulated banking and mortgage industries (2010, 26). Home foreclosures, tenant evictions, and homelessness are thus apt punishments. This perverted self-responsibility is organized and mobilized around discourses of crime and law enforcement, or what Giroux calls "the criminalization of social problems" (102).

Such criminalization involves the application and/or threat of violence by police officers, military personnel, and/or security guards. Refusing to vacate your foreclosed home, for instance, inevitably results in a confrontation with police. They will ask you to leave and if you say no, they will physically and if need be violently remove you. This demonstrates the neoliberal configuration of power. Power, defined as the ability to construct one's social world, is organized around the accumulation and protection of private wealth. All personal and social energies are filtered through this enterprise. Accepting and/or refusing this configuration of power leads to personal success or harm, respectively. Nothing demonstrates this better than the numerous families who have been kicked to the curb while the banking industry received $700 billion in taxpayer money (Johnson, 2008; McGreal, 2009).

Giroux's biggest concern is how this neoliberal-and-punish logic afflicts K–12 public education (2009). He argues that today's schools are increasingly viewed as "instruments of production and adjuncts to the corporation" and thus approached and protected as private property. Those who threaten that private property are treated as criminals to be managed and controlled by security forces. Police, probation officers, prosecutors, and various private security professionals are thus introduced into schooling systems. Such criminalization reconfigures pedagogical routines and practices. For instance, students who violated school rules were traditionally sent to a principal, guidance counselor, or another teacher. The students then explained and defended themselves to teachers and administrators who exercised their best judgment in issuing proper punishment. But such care and individualized discretion have been replaced by zero-tolerance standardization. This has led to record rates of expulsions, suspensions, citations, arrests, and referrals to juvenile courts. As Giroux (2009) states:

> Students who break even minor rules, such as pouring a glass of milk on another student or engaging in a school yard fight, have been removed from the normal school population, handed over to armed police, arrested, handcuffed, shoved into patrol cars, taken to jail, fingerprinted and subjected to harsh dictates of the juvenile and criminal justice system.
>
> How educators think about children through a vocabulary that has shifted from hope to punishment is evident in the effects of zero tolerance policies, which criminalize student behavior in ways that take an incalculable toll on their lives and their future. As the nationally syndicated journalist Ellen Goodman points out, zero tolerance has become a code word for a "quick and dirty way of kicking kids out" of school.

Back to the Campus

This same type of neoliberal-and-punishment logic motivates police repression of student protesting. On the one hand, basic, orderly protesting commonly signifies "traditional campus progressivism," which is then used for a college's promotional efforts. But on the other hand, more radical protesting is seen as a threat because it actually exposes the nature and distribution of campus power. Activists might reveal, for instance, the funding of corporate research, links between the school and military-industrial complex, exploitation of campus workers, use of sweatshop apparel, genetically modified foods used in the cafeteria, gentrification of surrounding neighborhoods, or general malcontent among students, faculty, and/or staff. College administrators are expected to safeguard the school from such public revelations. Any sign of unrest or negative press must be quarantined swiftly and forcefully. Failing to do so invites possible repercussions from boards of trustees, business partners, corporate sponsors, mass media, prestigious alumni, and even parents, students, and faculty who live by the neoliberal-and-punish logic.

Just imagine if the president of the hypothetical university described at the beginning of this chapter did not authorize police repression, but instead, publicly sided with the "Autonomous Students Insurrection." Suppose the president released the following statement:

> I have spoken with and listened to the concerns of the Autonomous Students Insurrection. After careful consideration, I have come to the conclusion that they are not only sincere but also accurate in their views and analyses concerning the conditions of our campus life as well as higher education as a whole. Given my sympathies, I agreed to establish a presidential committee that will meet weekly with ASI for the remainder of the school year. We will attempt to further explore and hopefully implement policies that will bring about a "Free Education for a Free People."

Such a statement would create a national uproar. The university president would be immediately removed and public relations specialists would rush in to refurbish the university's reputation as a neoliberal-and-punish patriot. Apologies, explanations, and assurances would be given by everyone—from board members to the city mayor. This demonstrates that the academic-corporate complex is more concerned with "brand value" than creating a more equitable, socially just society. Profit trumps social justice.

Developing an Oppositional Logic

There is no such thing as a politically neutral campus. Textbooks, course and degree offerings, styles of teaching, students' predispositions, the hiring and firing of faculty, administrative biases, and even the architectural designs of classrooms are imbued with political significance. If this is true, then those of us who oppose the repression of student protesting are morally obligated to construct a logic that challenges the neoliberal pedagogy. It is not enough to *say* that we disagree; we must actively create, champion, and apply a counter-logic.

This alternative logic must be based on critical democracy. It must be critical in order to safeguard us against the ills of status-quo operations. As Socrates said, "the unexamined life is not worth living," meaning that everything should be critically examined. This includes the rhetorical appeals to campus authorities and the excuses of "protecting the campus community." This counter-logic must also strive for democratic practice. The whole purpose is to empower people in every imaginable situation. This may not always be possible, but it is an ideal that guides critical inquiry and the fight for a better world.

The university, despite its problems, is in a unique position to develop and apply this critical democracy. That's because university life still provides space for experimental thought and discussion. To again quote Henry Giroux:

> As a vast array of public spheres . . . fall prey to corporate control . . . higher education becomes one of the few remaining sites where a society might question itself, where it might reflectively consider how lived realities measure against democratic practices and ideals. Universities thus provide the pedagogical conditions for existing and future generations to defend democratic principles and to incorporate them into their own understanding of what it means to define themselves as engaged citizens and socially responsible adults. (2010, 181)

I agree but also want to clarify Giroux's statement. Yes, the university can serve as a space for developing democratically engaged citizenship. But the very notion of democracy must be interrogated. Unlike the United States political system, a critically democratic logic is not reducible to representative politics—the practice of periodically voting for political leaders who then represent our wants and needs. This is the case in America where we vote for congressional and presidential representatives every two or four years. This system may be better than other forms of government, but it is still deficient. First, it distributes power in an upward fashion. We vote for others who then make decisions for us. Once elected into office, we have very little say in the policies that our representatives pass. Representative democracy thus maintains a hierarchy of power—representatives sit at the top while voters sit at the bottom. And second, representative politics actually negates democratic practice. Citizens are allowed to participate in and be counted by the representative system as long as they agree that the governing body can grant and/or relinquish their rights (and very citizenship) at any given time. We are therefore stripped of any individual, autonomous political power, and fundamentally prohibited from constructing the world according to our own free wills. At best, this is a perversion of democracy; at worst, it's the negation thereof.

In contrast, critical democracy tends toward a more direct, participatory democracy. It challenges the bureaucratic and formalized voting procedures, and proposes, instead, smaller, decentered assemblies and affiliations that better facilitate group decision-making processes. The goal is to expand rather than restrict political participation. This approach to democracy is inherently transformative, because it demands that we lessen our time with Facebook and reality TV and increase our participation in the creation and maintenance of our shared social world. Leisure time and entertainment are not negated, but instead, differently situated. Rather than being adjuncts to consumer capitalism, we become political creatures more willing and able to actively direct our personal and collective affairs.

It is important to reiterate that critical democracy *tends toward*—but is not synonymous with—participatory democracy. It starts but does not close with the thesis that participatory democracy is a viable alternative to neoliberalism and other socio-political pedagogies. This is because a critically democratic logic vigilantly resists any and all hegemonies that capture and enclose further thought and action. Even this logic of continuous interrogation is subject to scrutiny. Only through a willingness to be mistaken and corrected can we construct an inclusive, transformative pedagogy.

This type of critically democratic logic helps us resist such foregone conclusions as "student protesting is bad" and "police repression is sometimes necessary." More importantly, it remains open toward answering the question, "What is the nature and purpose of campus life?" Such openness is not relativism, but decentered and dialogical—different people respond through their own unique predispositions, and together we strive and struggle toward a never foreclosed horizon of ever-improving social relations. Any university that champions such a logic would be a revolutionary force within higher education. I believe that this is the type of university that we should strive to create, and if need be, defend: Free education for a free people!

Bibliography

Anderson, S., Cavanagh, J., Collins, C., Lapham, M., and Pizzigati, S. (2007). *Executive excess 2007: Staggering social cost of U.S. business leadership*. The Institute for Policy Studies and United for a Fair Economy. www.ips-dc.org/reports/executive_excess_2007

Aronowitz, S. (2000). *The knowledge factory: Dismantling the corporate university and creating true higher learning*. Boston: Beacon Press.

Bok, D. (2003). *Universities in the marketplace: The commercialization of higher education*. Princeton: Princeton University Press.

Bureau of Labor Statistics. (2010, 27 April). *College enrollment and work activity of 2009 high School graduates*. www.bls.gov/news.release/hsgec.nr0.htm

Del Signore, J. (2009, April 10). New School occupied again. *Gothamist Blog*. http://gothamist.com/2009/04/10/new_school_occupied_again.php

Domhoff, G. W.(2005, September; updated 2010, July). Wealth, income, and power. Author's website. http://sociology.ucsc.edu/whorulesamerica/power/wealth.html

Edu-factory Collective, The. (2009). *Towards a global autonomous university*. New York: Autonomedia.

Fleming, K., J. Doyle, and Nichols, A. (2009, April 10). New School occupation ends in 22 arrests. *New York Post*. www.nypost.com/p/news/regional/item_MbeyikAFfziXxqvLmEke0I;jsessionid=0C038FC1693FCFB87439012AC96244C2

Giroux, H. (2002). Neoliberalism, corporate culture, and the promise of higher education: The university as a democratic public sphere. *Harvard Educational Review*. 72.4: 425-63.

——. (2009, October 20). Schools and the pedagogy of punishment. *Truthout*. www.truth-out.org/10200910

——. (2010). *Politics hfter hope: Obama and the crisis of youth, race, and democracy*. Boulder, CO: Paradigm.

Giroux, H., and Myrsiades, K. eds. (2001). *Beyond the corporate university: Culture and pedagogy in the new millennium*. Lanham, MD: Rowman & Littlefield.

Hardt, M., and Negri, A. (2009). *Commonwealth*. Cambridge, MA: Belknap Press of Harvard University Press.

Johnson, C. (2008, July 11). Mortgage crisis is leaving children homeless. *Alternet*. www.alternet.org/story/91122

Johnson, J, and de Vise, D. (2010, March 4). Students protest cuts to higher-education funds. *Washington Post*. www.washingtonpost.com/wp-dyn/content/article/2010/03/04/AR2010030401307.html

Levine, A. (2001). The remaking of the American university. *Innovative Higher_Education*. 25.4: 253–267.

McGreal, C. (2009, November 12). US investigator accuses US of shameful neglect of homeless. *The Guardian*. www.guardian.co.uk/world/2009/nov/12/un-investigator-us-neglect-homeless

McKinely, J. (2010, March 4). California students protest education cuts. *New York Times*. www.nytimes.com/2010/03/05education/05protests.html

Moynihan, C., and Chan, S. (2009, April 10). Police arrest 22 at New School building. *New York Times*. http://cityroom.blogs.nytimes.com/2009/04/10/students-occupy-new-school-building-again.

National Education Association. (2004). *Higher education privatization*. NEA Higher Education Research Center. 10.2: 1-6. www.nea.org/home/34258.htm

New School in Exile. (n.d.), www.newschoolinexile.com

Occupy California. (n.d.) http://occupyca.wordpress.com

O'Leary, K. (2009, November 21). Tuition hikes: California and elsewhere. *Time*. www.time.com/time/nation/article/0,8599,1942041,00.html

Rae-Dupree, J. (2008, September 7). When academia puts profits ahead of wonder. *New York Times*. www.nytimes.com/2008/09/07/technology/07unbox.html?pagewanted=1&_r=1&ref=education

Starhawk. (1987). *Truth or dare: Encounters with power, authority, and mystery*. San Francisco: HarperCollins.

Washburn, J. (2005). *University Inc.: The corporate corruption of higher education*. New York: Basic Books.

Wertz, R. D. (2000, November). *Issues and concerns in the privatization and outsourcing of campus services in higher education*. National Center for the Study of Privatization in Education Teachers College, Columbia University. www.ncspe.org/readrel.php?set=pub&cat=47

Wharton, B. (2008, December 20). Notes from the The New School Occupation. *Dissident Voice*. http://dissidentvoice.org/2008/12/notes-from-the-the-new-school-occupation

Wollan, M., and Lewin, T. (2009, November 21). Students protest tuition increases. *New York Times*. www.nytimes.com/2009/11/21/us/21tuition.html?ref=education

• CHAPTER TWO •

Repression of Student Activism on College Campuses

Wesley Strong

Repression of free speech and assembly has a home on university campuses throughout the United States. Higher education is being corporatized, a process of privatizing public universities and colleges and developing corporate governance to centralize power in the hands of appointed administrators to limit or remove democratic rights from students, workers, and faculty. Repression of free speech and assembly play a key role in silencing opposition to these corporate reforms, all of which are designed to reduce the cost of education on the university and increase the cost of education on the student.

This is not unique to higher education, but is part of a much bigger crackdown on free speech and assembly in society. Various political changes took place following a series of economic reforms brought as a solution to a crisis in profitability beginning in the 1980s. These changes benefit the wealthiest class of citizens, the ruling class, those who own and operate the means of production, and mark a significant shift in free speech policies. This transformation could not be accomplished without suppressing their natural political opposition, manifested in the various forms of working-class political activity. This suppression targeted workers' unions, environmental groups, animal rights group, disability rights groups, feminist groups, anti-poverty groups, Black liberation groups, the global justice movement, the immigrants' rights movement, the anti-war movement, and various other working-class political formations.

These movements have faced heightened attacks since the Patriot Act was passed in 2001. The Patriot Act marked the beginning of an expansion of executive power. The Bush Administration (2000–2008) expanded intelligence and deterrence tactics, tapping an untold number of phone lines, placing police and cops in undercover operations inside peace and workers-rights organizations, rounding up and detaining Muslim Americans, and constricting political speech and assembly even further. The post–9/11 era is a testing ground for limitations on free speech and assembly, as well as other political rights, as the ruling class engages in wars overseas.

University administrations have been concerned about political speech and action since student and worker uprisings in the 1960s and 1970s. These uprisings disrupted the "business as usual" damaging rapport that administrators had inside of the upper echelons of society. Administrations aimed to rebuild this rapport, showing that they could have a "safe and orderly" campus. The growth of campus police began largely in the 1980s as administrators began to rebuild rapport with upper classes and began to execute neoliberal reforms in the education system.

The modern corporate university is based in this history, reflecting changes in society. The restrictions on free speech and assembly are directly tied to corporatization and free-speech limitations in the workplace. Higher education has become a ubiquitous requirement of most jobs, largely because of economic transformation away from manufacturing. This took place as the economy was transformed by neoliberal economic policies. The increase of jobs requiring a college education drove students into universities and drove universities into stronger competition over the best students. In order to compete and win, modern universities focus on building their brand and presenting students with the ideal university that will provide the best education.

This is part of a process best described as branding; the process through which university administrations develop an image of the school, in order to sell it to potential students, particularly the well educated upper class students that may bring some prestige to the university.. Universities focus on building their brand and presenting students with the image of the "ideal" university that is "safe" and a "haven from the world." In order to build and protect this image, university administrations limit political activism as it makes the administration look weak and unable to control the university, especially in the eyes of wealthy benefactors.

Many universities have been limiting political freedoms for years, if not decades; however, in points of crisis, these limitations are exposed. In this chapter, I will present several current cases that demonstrate these attacks. Secondly, I will analyze how these cases relate to the corporatization of education, the increased presence of police on campus, and the development of a culture of fear and silence.

The Methods of Repression

Academic Repression (Nocella, Best, and McLaren 2010) comes in many forms. They can be categorized by their content into four specific categories:

policy, non-violent police action, police violence, and legal action. Policies justify and defend police action and set the legal basis for repression of free speech and assembly. Non-violent police action includes surveillance of student and faculty organizations, harassment, and other tactics that do not involve physical violence. Violent actions, however uncommon, are used when prior actions fail to curtail the active expression of these core political rights. Legal action usually follows police actions and is used to sanction students for speech activity. These political tactics are used in tandem to target political speech, particularly controversial political speech.

Administrations began to establish a legal basis for the repression of free speech and assembly following the uprisings of the 1960s and 70s.. Laws were passed on a state-by-state basis allowing universities to establish their own police departments and hire commissioned officers. Universities established codes of conduct that include restrictions on speech and assembly. Many codes prohibit the use of hate speech or other broad categories such as "controversial" language or "offensive" speech. Universities enacted a series of other codes and policies targeting student organizations, creating more red tape for recognition and funding, and creating processes through which such groups can be sanctioned for the violation of these unconstitutional codes.

Policy

The Foundation for Individual Rights in Education (FIRE) produced several reports on the state of free speech and assembly on campus. FIRE recently published a report entitled *Spotlight on Speech Codes 2011*, which analyzes the state of free speech at over 300 major universities. The report finds that two-thirds of these universities have policies that violate or restrict free speech. Speech codes or codes of conduct may sometimes be constitutional as they are written, but enforcement is often selective and violates free speech and assembly of targeted groups.

Administrations broadened the definition of "inciting language" to include "offensive" or "provocative" speech. Inciting language is a recognized restriction; however, universities use it as a justification outside of its legal bounds. Incitement refers to language that encourages those who agree with the speech to commit immediate violence. Universities, in many cases, have used this as a legal basis to restrict speech that may incite anger from opposition. Offensive or provocative speech could include anything from racist slurs to political speeches, the latter of which is likely to be the target of enforcement.

Some administrations have also tried to charge students and student organizations a fee for extra security if they bring in controversial speakers or host controversial events. This is a violation of free speech according to the U.S. Supreme Court ruling in *Forsyth County v. Nationalist Movement* 505 U.S. 123 (1992), which states that "any policy imposing charges on speech, when those charges are based on an official's estimation of the likely disruption, necessarily requires an evaluation of the content of the message, and, therefore, both could and likely would be used to censor speech." The University of Oklahoma, Temple University, West Virginia University, and many others continue to implement similar policies.

Policies that establish free speech zones are in place on several campuses. These policies restrict political activity, demonstrations, and rallies to particular sections of a campus and during certain times of the day. FIRE provides example policies from Frostburg State University and the University of Central Arkansas, both of which designate certain areas of the campus as places to hold rallies and ban the use of all other areas for such activities. Students are often required to receive administrative approval for the use of "free speech zones" and in some cases several days ahead of time. Regulations are a form of prior restraint, in effect discouraging speech activity, particularly in situations that call for immediate response.

Many universities maintain policies requiring administrative approval for flyers, posters, and other materials. The overt intent is to make sure that posted materials are appropriate, though it certainly invites administrative officials to extend the limits of their mandate and prevent students from posting controversial material. Students were banned from posting fliers calling for the suspension of a University of California police officer following an incident at a protest during the 2010 Board of Regent's meeting. Jared Kemper, a University of California–San Francisco police officer, drew his gun on a group of student protesters following an incident brought on by a violent police response to the student assembly. The Center for Free Speech on Campus states that many universities have these policies in place, and some even explicitly say that the *content* of the flier must be approved before posting, like those in place at New Jersey Institute of Technology and Bridgewater State College.

Surveillance and Harassment

Surveillance and harassment are used by many administrations to curtail free speech and political assembly on campus. These "non-violent" tactics are designed to deter speech activity without using excessive force or physical violence. The purpose is to strike fear into political activists and make them second guess their role in political activity while also collecting information that will help police and the administration attack activists' credibility and counteract their political action.

Activists are under surveillance by campus, state, and in some cases by federal police forces. These targets are largely defined by their anti-war, workers' rights, environmental justice, or economic justice sentiment in almost every case. Police have even visited people at their homes to tell them that the police are watching them Police, in some cases, place undercover informants among activist groups to gather intelligence that will help prepare a police response. These efforts are part of a growing trend of "community policing" on campus. Informants are used in several cases, including recently throughout several University of California schools as well as in attacks on Palestine Solidarity, anti-war groups, and other groups that address issues of economic, environmental, or social justice. Campus, state, and federal agencies conduct regular surveillance of student groups on campus, particularly those that exercise political speech, as a part of the administrative reaction and management of student uprisings since the 1970s.

The American Civil Liberties Union (ACLU) catalogs the surveillance of groups in 33 states plus the District of Columbia in a 2010 report titled *Policing Free Speech.* The ACLU lists seventeen cases from university campuses out of a total of nearly one-hundred cases. These cases include military monitoring of counter-recruitment anti-war activity, undercover officers attending events on veganism, the interception and dissemination of private emails, preemptive arrest of protestors, among many other tactics.

Police and federal authorities engaged in active surveillance of email in several cases. Email surveillance tracked the activities of groups at University of California—Berkeley and Santa Cruz. E-mail surveillance was used at Central Connecticut State University, Colorado University, Georgia State University, and Evergreen College. Police infiltrated a private meeting of the Student Worker Coalition at the University of Washington. At Evergreen State College, police followed a student while he was in transit to a demonstration. Police were spotted in protests at the University of California-Berkeley shooting photos and gathering intelligence. Similar reports have come from other areas, most recently including University of California—Berkeley where police have been spotted inside of crowds taking photos for intelligence gathering.

Police harass activists for their political speech in several ways. They use house visits, questioning, and in one case, grand jury subpoenas. The FBI questioned members of the Idaho Progressive Student Alliance in 2004 following their boycott of Taco Bell. Grand jury subpoenas were issued in 2004 to students at Drake University following a seminar lead by the National Lawyers Guild on non-violent civil disobedience. The largest case is the harassment and persecution that is part of a targeted campaign to silence Palestine solidarity activists. The FBI raided several homes on September 24th 2010 in a coordinated raid at locations in Minneapolis and Chicago. Tracy Molm, a student and activist with Students for a Democratic Society, was one of the targets in Minneapolis. Molm previously visited Palestine as part of a non-violent mission. She helped organize a protest against the Republican National Convention in 2008 as part of the Twin Cities Anti-War Committee. These raids and subpoenas came on the back of a two-year undercover operation inside the Twin Cities Anti-War Committee, led by an FBI undercover operative.

Police engage in arrest activity that contributes to a culture of fear in activists communities. A student at the University of Arizona was arrested for using sidewalk chalk to advertise a protest. A Muslim student was arrested at Syracuse University for taking pictures related to a class assignment. At Penn State, a 21-year-old was arrested in his own backyard for taking a picture of police activity with his cell phone camera. Philip Chinn, a 22-year-old anti-war protestor and student at Evergreen State College was arrested on his way to an anti-war protest.

One aspect not covered in the ACLU report is the growing use of community policing campaigns. According to the Bureau of Justice Statistics 2004–2005 report on campus police, community policing programs are in place at over two-thirds of campuses. These campaigns included regular meetings with student group leaders, the development of emergency response plans, and a growth of security services on campus in an effort to address "security threats" and gather intelligence on student activities.

Students at the University of California–Davis recently discovered the university administration established an "Activism Response Team" to inform on the activities of student activists to the administration and police. Joshua Clover (2011), a professor of English Literature at UCD, discussed this in an article about surveillance on campus:

> Over the last year, the UC Davis administration has pursued an extensive program to place staffers in and around student-worker protest. They have done so not, as you might expect, to join in the struggle against indecent cuts and backdoor privatization, but to deliver surveillance on participants. This "Activism Response Team" was, for example, trained to "collaborate with police," and advised by university counsel on negotiating possible rights violations of those undergoing surveillance. When asked directly whether they were supplying information to the administration, ART members denied this.

Surveillance, harassment, and arrest curtail free speech and limit the extent and success of political action. These tactics are a form of prior restraint because they create a chilling effect on speech by establishing a precedent of harsh repercussions and harassment that create a sense of fear and apprehension among those looking to exercise political speech, especially when their content would be critical of the administration.

Police Violence

Police violence as a "crowd control" tactic has been in use on campus for several years. Several incidents of excessive force have been documented including the infamous "Don't tase me, Bro" incident and the case of the student who was tased in a library at UCLA for not producing his identification. Similar reports come out on a periodic basis; whether they involve police beating people at parties, the use of "less than lethal" weapons like tasers and tear gas, or dragging people in for arrest, police violence seems to be a growing trend.

Students began to coordinate their response to massive budget cuts and tuition increases in 2009. Student protested the University of California regents meeting in November of 2009 on the UCLA campus. Police used tasers on several students, sprayed mace on a crowd of students, and used metal barricades as battering rams against the crowd. A student had her hand crushed between a police baton and a metal barricade at a demonstration on the campus of UCB. Police used violent tactics towards student demonstrations on March 4th, 2010. In one such case, a student was beaten and dragged out of a peaceful demonstration by police as police confronted a peaceful march just outside the campus of University of California—Davis. A police officer drew his gun on a crowd of student protestors following an altercation initiated by police during protests at the 2010 University of California Regent's meetings.

Police respond with violence in many cases, and not just in California. Police attacked anti-war protestors outside of the presidential debate at Hofstra University in 2008. Police led horses into the non-violent crowd and one demonstrator was kicked in the face by a horse. Students at the University of Wisconsin—Milwaukee were assaulted, maced, and arrested by police as they tried to peacefully deliver petitions to the university president's office. Police violently broke up a demonstration for environmental justice at the University of Arizona—Tucson. Police ordered demonstrators to return to the designated "free speech zone" after students delivered a notice to a university official. As students began to return, police dragged several participants across the street, forced them to the ground, and arrested them.

Cases of police violence have a stronger chilling effect on speech, though they are less common as other forms of repression. Police violence can cause deep emotional harm to the victims. Violent attacks on activists contribute to a fear of taking political action, which is clearly a form of prior restraint. Police violence is defended by the courts citing principles of "safety," despite the chilling effect it has on speech.

Police violence serves two political purposes. First, it attacks and demoralizes activists. Second, it is used to criminalize dissent and speech. Police violence and arrest affect the emotional state of activists, particularly victims of police violence, and can create internal conflicts about strategy. Third, administrations support most acts of police violence and claim they are a justified response, painting activists as dangerous, violent, and out of control. These claims are often far from the truth, especially in cases where police violence is used against non-violent demonstrations like those listed above.

The preponderance of evidence shows that police are the central perpetrator of violence in these cases. Police physically repress free speech and freedom of assembly on campus and in larger society. Police take orders and direction from administrators who are guided by the ideology of the ruling class. For instance, Johannes Mesherle, the police officer who fatally shot Oscar Grant, shares the same protection as Jared Kemper, the UCPD officer who drew his gun on a crowd of protestors following an altercation. Police are used to enforce the political will of the upper classes, and as such are granted almost blanket immunity. In some cases, officers are reprimanded and jailed for unwanted activity, but in political cases, victims hardly if ever get any justice.

Police are the physical representation of the violence inherent in the undemocratic system, both at the university level and in society at large. The inherent violence derives from a system that steals future wages form young workers attending school, forcing them into a form of indentured servitude following graduation. This is an extension of the exploitation of labor in the workplace, as the ruling classes essentially steal value form workers by not paying them the full value of their labor. Ruling classes repress free speech in both the workplace and in education in order to maintain their hegemony and control over the means and distribution of production and education.

Legal

Most universities claim to have a functioning judicial system; however, these systems rarely meet the requirements of due process. Administrations use these systems to discourage political activity and contribute to a culture of fear on campus. Legal action is often used for political purposes by administrations, to limit the ability of activists to organize and hallenge the administrations policies of neoliberal austerity.

Three incidents occurred at the University of California–Irvine (UCI) in recent years. Members of the Muslim Student Association interrupted a speech by Michael Oren, Israeli Ambassador to the US, on Februrary 7th, 2011. The students continually interrupted his speech in an effort to shut down the event. The group was sanctioned by the university and several members were charged with criminal conspiracy by the Orange County District Attorney, possibly facing jail time. Nineteen UCI students face legal charges for a brief occupation of an administration building on February 24th, 2010. All students are facing between two and four misdemeanor counts including false imprisonment, obstruction of a public place, being a public nuisance, trespassing, disturbing the peace, and refusal to disperse. Several UCI students were detained by police in November of 2010 for chalking political messages on sidewalks. Sidewalk chalk does not damage property and is safe to use on many stone surfaces, including sidewalks. Courts have ruled sidewalk chalk as a form of vandalism, however, the enforcement is often selective and dependent on the content of speech, as was the case with this incident at UCI. Thirteen students were arrested following the 2011 regents' meeting in San Francisco and one was charged with assault on an officer following an incident where UCPD officer Jared Kemper drew his gun on a crowd of students.

FIRE reports over 20 open cases regarding due process on campus. George Washington University maintains a secret informer hotline. Charges can be made anonymously and the university can begin secret investigations based upon these anonymous charges. Three professors at Southwest College near San Diego were put on administrative leave after a group of students moved a peaceful protest from a free speech zone into a more public and visible location. The administration claims the professors incited the students to take this action. Hayden Barnes, a student at Valdosta State University in Georgia was expelled for protesting then-President Ronald Zaccari's plan to build parking garages on campus using $30 million in student fees. Barnes filed a lawsuit and has since been reinstated. Judicial proceedings, expulsions, suspensions, and other legal means are used to target free speech, particularly political speech.

Analysis

University administrators repress free speech in several ways. Their motives for repressing speech are based in the political and economic relationships this society is based on. Publicly, university administrators are concerned about the brand image of the university, but this is not the full motive behind repression. Administrators act as members of a managerial class, a sort of buffer class between workers and owners. Managers defend and forward the ideology of the ruling class in order to climb the social ladder, and hopefully progress into the ruling class. University branding, in this sense, is about appealing to a broad range of possible clients—i.e., students and their families—but more importantly about appealing to corporate donors, private interests, and the rich. Administrators, serving as a managerial class, are preparing students for the workforce by pushing them to conform to repressive conditions for political speech.

Administrations progress the corporate/private model of education as a political objective. This model limits or curtails democratic rights of the student, just as it does to workers. In *Schooling in Capitalist America* (1976), Samuel Bowles and Herbert Gintis argue that education is not a universally liberalizing system, but is in fact designed to replicate the next generation of workers and adapt them to the conditions of the modern economy. Bowles and Gintis argue that education is a system designed to "hide or justify the exploitative nature of the economy" (p. 13). The education system, in this sense, replicates the next generation of the working class and as such validates class divisions in society through a false sense of meritocracy. Education exists to replicate the next generation of the working class, clouding class division and wage labor exploitation behind a false system of meritocracy.

The writing and insights of Bowles and Gintis apply to higher education still today. The growth of policing on campus is a reflection of a growth of policing off campus, and the repression of free speech on campus is a reflection in the same manner. The growing corporatization of higher education brought greater levels of policing and repression of free speech to enforce neoliberal reforms and austerity. The corporatization of the university is part of a much larger set of changes—guided by neoliberal economic theory—for the purpose of dismantling the welfare state and restructuring society to be friendlier to private enterprise.

Neoliberal reforms began in the 1980s following a profitability crisis. The United States dominated an overwhelming majority of the world market during the post-war era as capitalist nations rebuilt following the devastation of the war. U.S. business brought in record profits—enough to sustain profit margins and follow a Keynesian welfare-state economic model. The profitability crisis of the late 1970s and the growth of international competition was resolved by liberating capital of taxes and tariffs, freeing it up for international investment and effectively beginning a process of recovering profit rates through layoffs, sweatshops, and a degradation of working conditions. Real wages—wages adjusted for inflation—have declined considerably since 1973. Corporations recognized greater profits by reducing expenditures on workplace safety, health insurance, and retirement/pension plans.

Higher education went through a similar transformation. Corporatization of the university led to massive cost cutting, which directly affected education outcomes. Students are continually being forced to pay higher tuition for the same or a lower quality of education. Student debt has become one of if not the biggest crisis facing students and youths today. For some people, the high price of education effectively prevents them from attaining higher education. This high level of exploitation creates profits for lenders—specifically banks, credit companies, and others—while it threatens the future of students and condemns them to decades of indentured servitude to pay creditors. Most of this transformation was incremental, however the economic crisis of 2008 expedited this process. University administrations used the crisis to forward the political goals of the ruling class, best represented by continued and expanded neoliberal reforms—cutting the university budget, raising tuition, breaking unions, outsourcing work—degrading the quality of education while raising the costs on students and workers.

The social concept of education in society at large is at odds with these political goals. Most working people see education as an avenue for social advancement, a path towards self-improvement. This liberal concept conflicts with the political agenda of the ruling class. The ruling class seeks to create the workforce they need, inherently restricting social advancement amongst the working class to the needs of the rich and powerful. This division, between owners and administrators in one camp and students and workers in the other, will inevitably lead to popular outrage against corporate/private reforms that attack the democratic rights of students and workers and centralize power in the hands of administrations. It is because of this division that administrators must repress free speech in order to advance their political aims. Corporation of free education *requires* repression of free speech be used as a political tool to threaten working class power and limit opposition to neoliberal reforms.

The current crisis presents as an historical moment in the fight against austerity and the progress of neoliberal reforms. Working people are beginning to organize against the political will of the ruling classes from California to Wisconsin, Occupy, and Chicago. Each of these movements have drastically changed the political climate in favor of working while also spurring responses from the ruling class in the form of greater repression of speech. Increased repression of free speech on campus is part of the ruling class political response to growing working class power in conflict with neoliberal reforms and austerity.

Free speech and assembly are central to the exercise of working class political power. Repression of free speech is an attack on the ability of working people to respond to political crises like neoliberalism, austerity, and the corporatization of higher education. We must understand that repression of political speech directly serves the goals of the ruling class and as such any movement against austerity, neoliberalism, and corporatization must also be a movement for free speech and assembly. We must confront repression with solidarity, creativity, and innovation to stay one step ahead of the ruling class and the agencies that enforce their politics.

Bibliography

Aggie TV. (2010). March 4th statewide day of Action at UC Davis. AggieTV. http://www.aggietv.org/?p=185

American Civil Liberties Union. (2010, June 29). Policing free speech: Police surveillance and obstruction of First Amendment-protected activity. American Civil Liberties Union. http://www.aclu.org/free-speech-national-security/policing-free-speech-police-surveillance-and-obstruction-first-amendme

Blakemore, B. (2006). Cops taser UCLA student. ABC News / US. http://abcnews.go.com/US/story?id=2662158&page=1

Bowles, S., and Gintis, H. (1976). *Schooling in capitalist America: Educational reform and the contradictions of economic life*. New York: Basic Books.

Bureau of Justice Statistics. (2008). Special report: Campus law enforcement 2004–05. Bureau of Justice Statistics. http://bjs.ojp.usdoj.gov/index.cfm?ty=pbdetail&iid=411

Center for Free Speech on Campus. (n.d.) Speech codes. Center for Free Speech on Campus. Center for Free Speech on Campus. http://www.campusspeech.org/page/cfs/speech-codes

Chan, N. (2011). UC Davis admin. monitored student activists involved in campus protests. *The Guardian*. http://www.ucsdguardian.org/news-and-features/features/item/22551-uc-davis-admin-monitored-student-activists-involved-in-campus-protests

Chiroux, M. (2008). A personal account by Matthis Chiroux of IVAW. *The World Can't Wait*. http://www.worldcantwait.net/index.php?option=com_content&task=view&id=5101&Itemid=220

Clover, J. (2011). Guest opinion. Private eyes: Real surveillance, false denials, from administration. *The California Aggie*. http://theaggie.org/article/2011/03/29/guest-opinion-private-eyes-real-surveillance-false-denials-from-administration

Coker, M. (2010, December 10). [UPDATE] 19 UC Irvine students and supporters face criminal charges for protesting. *OC Weekly*. http://blogs.ocweekly.com/navelgazing/2010/12/uc_irvine_protest_students_cha.php

——. (2011, March 11). [UPDATED] Irvine 11 charged with misdemeanors and face jail, probation, community service if convicted. *OC Weekly*. http://blogs.ocweekly.com/navelgazing/2011/03/michael_oren_irvine_11_arraign.php

Featherstone, R. (2008). Anti-Columbus Day protest disrupted by police violence. *Earth First Journal*. http://www.earthfirstjournal.org/article.php?id=104

Fight Back! News. (2010). Drop all charges from March 4 protest. *Fight Back! News: News and Views From the Peoples' Struggle*.http://www.fightbacknews.org/2010/3/7/drop-all-charges-march-4-protest

Foundation for Individual Rights in Education. (2011a). Southwestern college: Professors banned from campus after students stray beyond 'Free Speech Patio.'" FIRE. http://thefire.org/case/806.html

——. (2011b). Due process. *FIRE*. http://thefire.org/cases/dueprocess/

——. (2011c). George Washington University: Secret informer line. FIRE. http://thefire.org/case/38.html

——. (2011d). Spotlight on speech codes reports. FIRE. http://thefire.org/code/speechcodereport/

——. (2011e). Valdosta State University: Student expelled for peacefully protesting parking garages. FIRE. http://thefire.org/case/751.html

——. (2011f). "FIRE's guide to free speech on campus. FIRE. http://thefire.org/article/5063.html

Hesse, M. (2007). Aiming to agitate, Florida student got a shock. *Washington Post*. http://www.washingtonpost.com/wp-dyn/content/article/2007/09/18/AR2007091802115.html.

Nocella, A., Best, S., and McLaren, P. (2010). *Academic repression: Reflections from the academic industrial complex*. Oakland, CA: AK Press.

Schultz, J. (2002). Court rules sidewalk chalking not free speech. *Santa Cruz Independent Media Center*. http://santacruz.indymedia.org/newswire/display/2020/index.php).

Strong, W. (2009). Goin'back to Cali. *New Student Left Review*. http://nslreview.net/goin-back-to-cali

The university belongs to those who use it. (2010). Protest at Regents' meeting marked by indiscriminate police violence. The University Belongs to Those Who Use It. http://thosewhouseit.wordpress.com/2010/11/17/protest-at-regents-meeting-marked-by-indiscriminate-violence/

Woo, M. (2010). 9 UC Irvine students detained for "chalking." OC *Weekly*. http://blogs.ocweekly.com/navelgazing/2010/11/9_uci_students_detained_for_ch.php

• CHAPTER THREE •

Policing College Campuses

Race, Social Control, and the Securitizing of College Campuses

Daniel White Hodge

I remember it was a late fall semester on a private Southern California college campus. I had just finished teaching a four-hour course on intercultural communication. It was a draining class, as many of my students had never heard of issues such as race relations, mirco-aggression, and/or color-blind racism. Therefore I was anxiously anticipating my quiet ride home listening to Donna Summer. I had parked my car in the faculty parking lot adjacent to the lecture hall I was teaching in. As I approached my vehicle, I felt as though I was being followed and watched. I stopped, looked around, but seemed to be the only person in the parking lot. Then I heard it; the squelch from a radio and the low talk that preceded it. I was being followed by campus security. Not paying it much mind, I made my way to my vehicle when two squad cars swiftly arrived on the scene. The four campus officers exited quickly and ecstatically told me to stay where I was, take my hands out of my pocket, and stay still. It was then that I noticed my "tracker" who had been following me; he came up and started looking me over and asked me questions such as "where were you headed," "where had you been," "why were you on campus," and "what were your whereabouts at a particular time." It seems some student—who would later be revealed to be a White female student—felt "unsafe" and reported a "suspicious" Black guy walking around campus; that "suspicious Black guy" was presumed to be me, and I was now in the clutches of this campus's finest.

After about a 30-minute delay and a barrage of questions, I was released to "be on my way." Once the officers figured out I was a faculty member and that I was in fact teaching that night—although one officer thought I had presented a fake identification card because rapists typically carry nine textbooks in their bags along with fake identification—I was told "We have to follow up on every case, sir. One can never be too sure. Sorry for the inconvenience. We strive for a safe campus." How could one lone call generate such a response from campus safety? How could it be that with my credentials and time in at this particular campus, I was still considered a "threat"? Moreover, what if I had not had the correct "identification" at the time? What if I was a Black student headed out to buy a snack at the local convenience store and not brought all my identification? Was race really a factor? And what is the obsession in this country with safety and feeling "safe"?

Later that same year on the same college campus, another White female student said she had been abducted, sodomized, and beaten by one Black male and two Latino males. She made a full report, police launched an investigation, and neighboring schools were put on alert; the community, once again, felt "unsafe" around these "aggressive gangs." The young woman later confessed she had made the story up and the results of the police investigation supported her confession. Works out? Not before dozens of young Black and Brown males were arrested, detained, and questioned for hours—in some cases, literally minutes after this young woman made the report. The damage had been done. No charges, to my knowledge, were ever brought against the young woman for falsifying a police report. She was admitted into the school's "counseling" program for observation.

In another incident, on a community college campus, a new Black student organization was placed on "watch" as several anonymous phone calls were placed to the president of that school saying they were organizing riots and "mayhem" while the main student body organization—predominantly White—had been involved with protests of the Iraq war, protesting budget cuts, strongly criticizing school administration for their lack in connection to students, and were involved in several scuffles with campus safety—none of which drew any concern from the campus.

In yet another event, White students on a Cal State Los Angeles campus threw bottles at police, broke several campus windows, and vandalized several administrators vehicles, and only one arrest was made, with the charges later being dropped for "lack of evidence." While Black and Brown students across the way at another Cal State campus were pepper sprayed, beaten, and arrested as a result of a rally against police brutality.

What is the cause of this? Are campus safety officers racists? Is there truly a racial bias that exists on college campuses around our country? Do White students get treated differently on college campuses than Black, Brown, or even Asian students?

The reality and honest answer to all of these questions is yes. Yes, many campus safety officers are inexperienced, ill-trained, and have never been through any racial, cultural, or gender sensitivity training (Apple, 2008; Peak, Barthe, and Garcia, 2008; Tillyer and Klahm, 2011). Moreover, many college campus police are private law enforcement agencies—which typically fall out of the state and federal mandates for law enforcement—deepening the issue of race. When looked at carefully, White male police officers dominate law enforcement agencies (Cordner and Cordner, 2011; Schulz, 2003). This phenomenon is only exaggerated further on campus police units (e.g., Brown and Benedict, 2002).

To add insult to injury, media representations of ethnic minorities—especially Black and Brown males—are embellished on television shows like *COPS*, *Law & Order*, and *CSI* along with talk shows such as *Montel Williams*, *Ricki Lake*, *Sally Jessy Raphael*, *Jerry Springer*, and *The Maury Povich Show*, which portray Blacks as amoral, buffoons, aggressive, sassy single mothers, arrogant absent fathers, and potentially violent individuals (Russell, 1998). These perceptions carry over into law enforcement agencies and interactions with ethnic minority students on campuses across the country. When Blacks are portrayed in the media, print, and in film as loud, violent, and aggressors, they are inevitably targeted when a "crime" or "safety" issue arises on college campuses. Add this ethos to the sentiment that many White students feel Black students are merely on campus because of programs like affirmative action and quotas, and you have even more animosity and fuel for the "fire" (Leo and Davis, 2009)

Violence, rage, danger, loudness, arrogance, troublemakers, and rapists are all descriptive terms assigned to Black and Brown ethnic minorities. When these mantras enter a college campus, they are exacerbated further on all-White campuses that have little to no understanding of intercultural communication and/or the history of race relations in this country. However, when a "problem" does arise, Black and Brown students are some of the first to be blamed, targeted, and arrested by campus police.

These narratives that I have listed here—which are mere scratches on the surface of what many Black and Brown students face on college campuses across America—demonstrates the relentless social inequality that continues to come in the form of racism. Further, they demonstrate that policing on college campuses is biased and needs major overhauling in order to prevent further problems. Because as Kumasi Brown so eloquently puts it in the documentary *Bloods and Crips: Made in America* (Peralta):

> In a free society, in a free country; I'm walking down the street police got the nerve to ask me, "Where you going?" "What you doing?" Ain't none of your damn business where I'm going. Ain't none of your damn business where I came from. Do you do that shit up in the suburbs? You stop anybody else in this society and ask why do you exist? But you got the nerve to ask me all day, everyday? Now, what do you think that does to me psychologically? What does that tell me, what message am I being fed? Every day. See, he doesn't understand something; he is feeding me a spoon full of hate; every day.

> Everyday that's my diet; a spoon full of hatred. Now it's just a matter of time before I'm going to go off...

The ensuing after effects of this police state can, and will, be costly for the future.

This chapter is concerned with illuminating and challenging the police state mentality that exists on many college campuses across the United States. Further, I make the argument that many college campus police units are not prepared—either culturally or racially—to handle the diverse group of students entering college campuses over the next two decades, and they stereotype, based on media and engrained stereotypes, ethnic minority students as violent, aggressors, and potential threats on campus. My narrative, along with countless others, are just a scratch on the racially charged iceberg that exists within campus police units on college campuses, and the discussion is imperative as we move forward.

The Color of Crime

When I asked students in my introductory sociology class, who were predominantly Black, how many of them had been stopped, detained, and or questioned by campus safety, almost the entire class raised their hands. Upon further probing, I found that chiefly the Black males were the ones targeted and at least one out of every two of them had been detained, later released, for more than two hours over what they described as petty and irrelevant issues. Is this just a fluke and possibly a one-time deal? The following semester, I posed the same question to a more "diverse" set of students and the results were the same in regards to the Black and Brown students—predominately males—being detained by campus security.

This should come as no surprise when news stations, certain "reality" shows, biased statistics, and television shows focus on violent street crime typically illustrating Black males as the cause of this violence. Moreover, shows like *COPS* elucidate Black and Brown males as potential muggers and rapists. This is contrasted with images of Blacks—in particular—as lawyers, doctors, and some police officers. Yet, the dominating stereotype is still that of the violent uncivilized thug (Russell, 1998). Further, this fictionalized reality puts the public, especially law enforcement, on edge and filled with mental tapes of Black and Brown males being violent ruffians. Katheryn Russell reminds us that "Television news, with its focus on violent street crime, also fuels the stereotype of Black criminality. Many local news programs lead off with a crime story...with a [Black male] prodigy hunched over shielding his face from the camera, being escorted away in handcuffs by the police" (Russell, 1998).

Thus, for many, these types of images are overwhelmingly difficult to overlook and ignore. It makes the law officer feel alert and suspicious of "groups of Black men" particularly at night (e.g., Rydberg and Terrill, 2010; Tillyer and Klahm, 2011). This onslaught causes many to incorrectly judge Black and Brown men and unjustly conclude that these types of men—especially if there are tattoos, bald heads, and any type of "urban wear" associated with them—are criminals.

On October 8, 1995, Kevin Merida of the *Washington Post* published a three-part series on "A Distorted Image of Minorities." The study reported the results of interviews with 2,000 adults. Latinos, Asians, African Americans, and Whites were surveyed and polled to examine and study how information and misinformation shapes public thinking on race. The participants were asked to estimate the percentage of Blacks in the United States, the correct answer at the time being 12 percent. However, each ethnic group—including African Americans—overestimated that the United States, on average, was well over 25 percent and in one case, one White female estimated that African Americans were over 50 percent of the U.S. population. Once she was told that the number was closer to 12 percent, she responded by saying, "Census must not have been done correctly.... It's only common sense. They [Blacks] have very, very large families" (Merida, 1995). When the crime element is added in, the perception goes much higher.

To contrast that, Asians, Latinos, and Whites assumed African Americans fared better "in general" to Whites and that "Blacks were better off" than Whites in the United States So what accounts for these types of distorted images? Once again, Katheryn Russell (1998) asserts that, "At core, the presentation of Blackness is one of contradiction. Images of Black deviance are promoted alongside images of Black achievement—sometimes they overlap." To compound this problem further, in Hollywood, Blacks are typically and often easily given roles of gang members, street hoodlums, and/or pimps who are violent, mean, and very loud. Keep in mind all the great roles Denzel Washington has played, but the role that landed him the Oscar was for the thug rouge cop he played in the film *Training Day*. Russell states:

> The picture of Black success is partly presented in the form of fictional media portrayals of Black and White friendships. On the movie screen for instance, Blacks and Whites are paired as partners-in-crime, blood brothers, coworkers, and friends.[6] These images, however, do not effectively counter negative Black images. This is because sometimes underlying these Black and White pairings and images of Black success are subtle messages of Black deviance. Television and the movies frequently portray Black achievement as a "reform" case. On the television show *Designing Women*, the Black character, played by Anthony Bouvier, was an ex-convict, hired as the gofer for four affluent Southern White women. On *Doogie Howser, M.D.*, the Black character was a rehabilitated stick-up kid. In the movie *The Shawshank Redemption*, two convicts, one Black, one White, become fast friends.

Notably, the Black character, Red, was guilty of murder, while the White character, Andy, had been wrongly convicted of murder.

Law-abiding young Black men are often overlooked for the simple reason that oftentimes their voice is silenced in the literature and research. Further, the media socially constructs a crime-filled race of Black men waiting to "pounce" on their latest victims (Merida, 1995; Peak et al., 2008; Peart, 2011). In 1996, The Council on Crime in America reported that "the generation of urban males approaching their teens is a ticking time bomb, a national wolf pack of conditioned super-predators." Although the author of this report, John DeLulio, later recanted his statement, this "super-predator" concept had been unleashed, creating a web of legislation in a time when crime was actually on a decline (Brown and Benedict, 2002; Peak et al., 2008; Russell, 1998). Combine this "super-predator" ideology with race, and you have an extremely deadly mix.

Varying Responses to "Safety"

In late 2011, in response to the firing of Coach Joe Paterno, a largely White male crowd of Penn State students gathered in a commons area to vent their frustration over Joe. What started out in a "civil" manner quickly escalated into a full riot. Cars were overturned, trashcans were thrown through windows, trash bins were lit on fire, and the police were called in. While there were some arrests, the media downplayed the riot and issued mass sympathy for the "hard times" the school was facing. One Fox News reporter even went as far as to say "these poor students didn't have any other way to vent their frustration." Karen Schrock in *Scientific American* wrote an essay on why students "deify" Joe Paterno, and many of the students arrested were later released on minor charges (Schrock, 2011). No expulsions were reported, and the media spent very little time on the entire issue. By the next week, little to nothing was reported.

Yet, when students on Berkeley, Cal State Los Angeles, UCLA, and San Francisco State campuses were "frustrated" over the death of Oscar Grant, who was shot by a B.A.R.T. transit cop at point-blank range while handcuffed, they rallied on campus common grounds and within minutes SWAT, campus safety, and local police were called in—some in full riot gear—to squash what had been termed a full-scale "riot." The media has much less sympathy for "these students" who were shown to be predominantly Black and "angry." In fact, many were not only cited for disturbances of the peace, but were also disciplined by their colleges for "unruly behavior."

Hence, Hamilton Nolan (2011) asks the poignant question "What if everyone involved in this Penn State scandal were Black?" What if? In fact, what if

Penn State students were Black and rioting over a man accused of child molestation—who was also Black—and set cars, classrooms, and trash bins on fire?"

Policing campuses has less to do with actual "policing" and more to do with the culture of fear within racial constructs of crime (cf. Sutherland, 2001). The *safetyization* of college campuses has much to do with the ongoing fear that "America is changing" and "we must take it back" mantras felt by many Whites who see a social landscape ethnically changing over the next 25 years. Policing campuses is, in one sense, a part of that backlash and last effort to control power among White hegemonic structures. Thus, Blacks on campus, while in small number and domesticated in stature, do not present an immediate danger, but in greater numbers and when agitated and loud are now a threat and a cause for alarm.

More importantly, when issues such as the Oscar Grant case, Troy Davis, and even events similar to the Rodney King beating arise on college campuses, it is seen with disdain and fear from White students—whose parents still control much of the power and endowments sent to many tier-one institutions across the U.S. And part of the tacit function of many campus safety units on college campuses is to keep that "safety" structurally controlled. One college president even told me once, after some very wealthy alumni were offended by LGBT students wearing pride colors on their graduation gowns, "How long do we keep this 'diversity thing' going before it's just bad for the school." In yet other cases, a student who was a campus safety officer informed me that campus safety was actually taught to go after and monitor "Black and Brown troublemakers" and to keep other students safe from "them."

In still other cases, campus safety has harassed and at times detained Blacks simply passing through the campus; another former campus officer told me that they would almost always detain and hold a Black male who looked "suspicious" if seen on campus. At one point, this was such a problem that when I held events on campus that I knew "outsider" Black males would attend, I would take their pictures over to the campus safety office just to avoid problems—often time the changing guard would not inform the next shift, and these males would be stopped and questioned even still.

Safetyitizing a campus does not happen merely within the campus safety office. It is a structural event. At several board meetings I was privileged to attend on prestigious private colleges across Southern California, wealthy and powerful alumni made it a point to discuss how the campus would keep their child safe. Now, there is no issue taken here with a parent simply wanting to protect his or her child. However, the clear subtext was that the campus was changing "color" too "rapidly" for some board members and the ensuing climate on campus was "changing" much for the "worse" as they put it. One person even looked over at me and even asked why I was there—as if to say "Your kind doesn't belong here." The presidents of these colleges assured them that "every effort is being taken to secure these campuses" and that "safety is our number one priority." One can read between the lines at these mainly White board meetings what that might in fact mean for ethnic minorities; or, even faculty as myself leaving campus late at night.

Campus police, sometimes managed by former police officers, or even worse, individuals who never made it to or flunked out of the police academy, tend to view the world in a very black-and-white lens; you either did it or you did not; you are either guilty or not. Therefore, the issue of racism within structures and systems is overlooked and even ignored. As one campus officer who was a captain put it when discussing the issue of racism in his department, "Those issues are long behind us. They are long over. Those who keep dwelling on those issues are typically the ones causing trouble. Our officers are color blind. We're here to protect the campus."

To once again compound the issue even further, Weitzer and Tuch prompt us that:

> Americans are ambivalent about the police. We are fascinated by them, as evident in the popularity and proliferation of police shows on television—from dramas such as *NYPD Blue* and *Law &Order* to reality programs such as *COPS*. These shows typically present the police in a sympathetic light, even when they act aggressively or improperly against citizens. Programs such as *COPS* appear to present the "reality" of everyday patrolling, but with a camera crew present the featured officers are obviously on their best behavior. (1)

In fact, many in the public—including many upper-middle-class Blacks and Latinos—believe that the police are "doing their job" keeping criminals out of the their homes and neighborhoods (Weitzer and Tuch, 2006). The public, when given the opportunity, will group-think quite easily with the worldview that says "you do the crime, you do the time." So we can expect no "grace" from a public who views bad/good in these terms. Most of the public is unaware of the gross inequality that exists between street petty crime and white-collar crime. We are reminded that as of yet, there have been no charges brought against A.I.G., Country Wide Banking, Chase, Citibank, Wells Fargo, and Bank of America CEOs who aided in the great depression of 2007; moreover the executives at A.I.G. who "took" half a million dollars of bailout money and spent it on spas and "pampering" have still not been brought to trial for theft.

It is almost impossible to overlook the police state that quietly sits on every college campus across the United States. This police state exists to maintain social status quo and to keep "Them" in place while keeping the "peace" on campus. With the privatization of many social institutions, campus safety officers—many of whom are licensed to carry weapons and in some cases full automatic weapons, as happened following the Virginia Tech shootings—are a growing giant who imposes a serious threat to the welfare and well-being of not just ethnic minority students, but also poor White students. Private companies are beginning to dominate the landscape of the United States; they are more profitable, quicker, and oftentimes fall outside of the local, state, and federal regulations. This, once again, poses a serious hazard to how we "police" our college campuses.

Concluding Thoughts

Campus safety officers have little to no training on issues related to cultural and racial diversity (Brown and Benedict, 2002; Leo and Davis, 2009). Consequently, it should not be surprising when racism and general intercultural ignorance are found in these police units. Moreover, it should come as no surprise that the individuals who tend to join these types of organizations could potentially, themselves, have issues with ethnic minorities (Brown and Benedict, 2002; Parrotta and Thompson, 2011; Peart, 2011; Tillyer and Klahm, 2011), therefore adding even more fuel to the ongoing bonfire of this campus police state.

In a moving *New York Times* article, Nicholas Peart (2011) narrates what it is like to be frisked, searched, detained, and humiliated by the NYPD:

> For young people in my neighborhood, getting stopped and frisked is a rite of passage. We expect the police to jump us at any moment. We know the rules: don't run and don't try to explain, because speaking up for yourself might get you arrested or worse. And we all feel the same way —degraded, harassed, violated and criminalized because we're black or Latino. Have I been stopped more than the average young black person? I don't know, but I look like a zillion other people on the street. And we're all just trying to live our lives.

This is the sad reality for many young Black and Brown men.

Further, this college police state is a symptom of an even larger problem of the ongoing (1) prison industrial complex, (2) police aggression against ethnic minorities, and (3) growth of police brutality. Campus safety is merely the sub-arm of these serious phenomena.

This essay has been concerned with revealing and challenging the police-state mentality that exists on many college campuses across the United States. The police state, which we can find on almost every college campus, is an ongoing threat to ethnic minority students who merely wish to gain an education. I have used narrative and ethno-narrative to give illustrations of this problem. However, further research is needed in this area and an empirical study needs to be engaged in order to better understand and deal with the police state on college campuses. Because as Richard Delgado so importantly argues, "No one focuses on White crime or sees it as a problem. In fact, the very category 'White crime' sounds funny, like some sort of debater's trick." This is, hopefully, the first of many studies that illuminate this issue.

Bibliography

Apple, M. W. (2008). Can schooling contribute to a more just society? *Education, Citizenship and Social Justice, 3*(3), 239–261.

Brown, B., and Benedict, W. R. (2002). Perceptions of the police: Past findings, methodological issues, conceptual issues and policy implications. *Policing: An International Journal of Police Strategies & Management, 25*(3), 543–580.

Cordner, G., and Cordner, A. (2011). Stuck on a plateau? *Police Quarterly, 14*(3), 207–226.

Leo, R. A., and Davis, D. (2009). From false confession to wrongful conviction: Seven psychological processes. *Journal of Psychiatry and Law.*

Merida, K. (1995, October 8). A distorted image of minorities. *Washington Post.*

Nolan, H. (2011). What if everyone involved in this Penn State scandal were Black? http://gawker.com/5858232/what-if-everyone-involved-in-this-penn-state

Parrotta, K. L., and Thompson, G. H. (2011). Sociology of the prison classroom. *Teaching Sociology, 39*(2), 165–178.

Peak, K. J., Barthe, E. P., and Garcia, A. (2008). Campus policing in America. *Police Quarterly, 11*(2), 239–260.

Peart, N. K. (2011, December 17). Young, Black and frisked by the N.Y.P.D.: Why is the N.Y.P.D. after Me? *New York Times*, p. 3. http://www.nytimes.com/2011/12/18/opinion/sunday/young-black-and-frisked-by-the-nypd.html?pagewanted=all&_moc.semityn.www

Peralta, S. (Writer). (2008). *Crips and Bloods: Made in America* [DVD]. USA: Verso Entertainment.

Russell, K. K. (1998). *The Color of Crime* (Kindle ed.). New York: New York University Press.

Rydberg, J., and Terrill, W. (2010). The effect of higher education on police behavior. *Police Quarterly, 13*(1), 92–120.

Schrock, K. (2011). Why Penn State students rioted—They deify Joe Paterno. *Scientific American.* http://www.scientificamerican.com/article.cfm?id=penn-state-students-ri

Schulz, D. M. (2003). Women police chiefs: A statistical profile. *Police Quarterly, 6*(3), 330–345.

Sutherland, E. H. (2001). Is "white-collar crime"crime? In H. N. Pontell and S. M. Rosoff (Eds.), *Social Deviance: Readings in Theory & Research.* New York: McGraw-Hill.

Tillyer, R., and Klahm, C. (2011). Searching for contraband: Assessing the use of discretion by police officers. *Police Quarterly, 14*(2), 166–185.

Weitzer, R., and Tuch, S. A. (2006). *Race and policing in America: Conflict and reform* (Kindle ed.). New York: Cambridge University Press.

• CHAPTER FOUR •

Policed Pedagogy

Controlling and Dominating Classrooms, Curriculum, and Courses

Kim Socha

In fall 2011, I sat on a panel at a Minneapolis bookstore based upon the anthology *Academic Repression: Reflections on the Academic Industrial Complex* (Nocella, Best, and McLaren, 2010). My co-speakers were Anthony Nocella, II, one of the text's editors and authors, and David Pellow, Don Martindale Endowed Chair of Sociology at the University of Minnesota. I wondered what I, a community college instructor, could add to the conversation on academic repression and, by necessity, academic freedom. After all, very few community colleges show up on the American Association of University Professor's censure list (Franke, 2006, para. 1). However, as community college faculty teach about 50 percent of all first-year college students in the United States, I decided that we need to be heard more regularly on issues of pedagogical repression (Wilson, 2010, para. 8). As I prepared for this event, I realized just how much repression at community colleges manifests as policed pedagogy, meaning that instructors at two-year schools are intra- and inter-departmentally supervised, often to the detriment of the learning environment.

Before exploring policed pedagogy at community colleges, it is important to define the term as I use it herein. Policed pedagogy should not be confused with repressive and controlled pedagogy, with the cause of policed pedagogy being external—not at the teacher's will and often against her desires. Repressive pedagogy and controlled pedagogy, on the other hand, are embraced and supported by the educator. Nocella (2006) writes that "repressive pedagogy is designed and used by military, fascist regimes, and dictatorial militia groups in hopes to have non-questioning agenda-driven members, which promote or conduct their actions under false-truths" (para. 31). Repressive pedagogy teaches repression through a strategic propaganda campaign. Controlled pedagogy is teaching from an authoritarian and dominating approach, while not promoting a particular ideology or mission.

Policed pedagogy, a specific type of academic repression, refers to teaching under external repressive and controlling conditions and influences. Policed pedagogy can take one of three forms: (1) by administration dictating curriculum, lesson plans, methods, physical place of instruction (i.e., online, in the classroom, or hybrid) and materials used; (2) by law enforcement, military or other armed agencies through physical intimidation and coercion; and (3) through propaganda campaigns seeking to instill fear and stigmatization in faculty, staff, and/or students.

Community colleges are ripe for examples of policed pedagogy, for they have historically had to rely on contingent faculty, frequently hiring people who don't view community college teaching as a lifetime profession but as a weigh station to positions at four-year colleges and universities. This has placed departments in a position to claim that they *must* police what and how those contingent faculty teach, since they are often a transient population who must conform to a particular school or school system's curriculum. In some cases, part-time faculty cannot even choose their own textbooks. It is almost always the career adult-educators who write the departmental syllabi from which contingency faculty are forced to teach. This is a prime example of Michael Apple's (n.d.) concept of "deskilling": "Instead of teachers having the time and the skill to do their own curriculum planning and deliberation, they become the isolated executors of someone else's plans, procedures, and evaluative mechanisms" (p. 33). Usually, part-time faculty become the "executors" of full-time faculty's "plans, procedures, and evaluative mechanisms." This is true not only for contingent faculty, but also for new full-time faculty who are given a set of ossified guidelines determined by a small departmental committee.

Fortunately, there is freedom within those guidelines, with many instructors—both part- and full-time, tenured and non-tenured—using their creativity and ingenuity to modify common course outlines. However, this freedom is also being challenged at community colleges, including within my own school system of 31 colleges and universities. In the article "New MnSCU [Minnesota State Colleges and Universities] Chancellor Promises to Shake Things Up" (Associated Press, 2011), Chancellor Steven Rosenstone said he would consider instituting a "single online English 101 class system-wide" (2011, para. 4). While he didn't elaborate, I and others in my English department are wary of the implications that every English instructor who teaches one of the most widely taken courses in the MnSCU system will not only be required to teach it online, but they will also be forced to accept whatever standardization is imbricated in his plans. This proposal likely ensures that a single bureaucracy of English teachers will make determinations for how *all* instructors run their courses within all 31 schools.

In sum, policing has long been endemic to community college teaching, and there are signs of its intensification, my chancellor's proposition being one example. With the increasing popularity of community colleges as starting and ending points for those seeking associate's *and* bachelor's degrees (some community colleges now offer four-year degrees), these issues of policing are more important to consider than ever before as they affect more citizens trying to obtain college degrees. Indubitably, community colleges are on the rise: "Total fall enrollment in community colleges increased from 5.7 million in 2000 to 6.2 million in 2005 and 7.1 million in 2009. In 2009, 2.9 million students (41% of the total) were enrolled full-time" (Baum, Little, and Payea, 2011, p. 3). And they are on the rise for practical reasons. Baum, Little and Payea (2011) report that a full-time student at a two-year school saves about $5,000 per year compared to if s/he had attended a four-year public institution, and "significantly" more money than if s/he enrolled in a "private for-profit [or] nonprofit institution"; further, tuition growth rates are lower at two-year colleges (pp. 1–2).

The policing of college and university campuses clearly falls under the topic of academic repression. While other chapters in this text focus specifically on strategies and tactics by campus safety, peace officers, and police, this chapter focuses on what administration is doing to control and dominate curriculum and education, which I refer to as policed pedagogy: that which prevents, silences, represses, and controls teachers from being free to do their jobs as they best see fit. Thus, this chapter is dedicated to examining and critiquing two dilemmas in higher education: a culture of fear that strengthens as more colleges rely on part-time faculty and the issue of curriculum, lesson plans, and teaching practices being dictated through policed pedagogy at community colleges.

Higher Education and the Culture of Fear

In *The Emerging Police State: Resisting Illegitimate Authority* (2004), radical activist and attorney William Kunstler's 1995 address to the State University of New York Buffalo's School of Architecture and Planning notes the power that fear has in making individuals acquiesce to oppressive regimes and power structures:

> We've come to the point, I guess, where we fear so much—crime in the streets, bombings, domestic terrorism, and the like—that we are virtually willing to countenance giving up of rights because we think it will safeguard us in our

> daily lives . . . . We are succumbing, in a way, and I don't make the analogy too close, to what the German people did when the Third Reich began to plant its foot on human rights in Germany. (p. 24)

This fear that Kunstler describes found its way into higher education after September 11, 2011. Since that event, higher education changed drastically, shifting priorities from critical thinking and the freedom of intellectual discovery to aiding in the development of the corporate university and academic industrial complex (Nocella, Best, and McLaren, 2006, p. 13; Giroux, 2006, p. 94). This is especially true, of course, when professors dare to question the legitimacy of the U.S. government's official accounts of the events of 9/11. For instance, scholar Kevin Barrett reports becoming the "target of a neoconservative corporate-media smear campaign aimed at dissuading professors from speaking publicly on 9/11" (n.d., para. 3). This new academia that Barrett bemoans has two dominate goals: (1) to develop partnerships and contracts with corporations on their research and to train students to enter the workforce; and (2) to promote a militaristic conservative agenda by funding faculty and student research, program development, and law enforcement and military partnerships on national security and counterterrorism that supports U.S. imperialism and to report, silence, and marginalize any and all possible threats to this agenda in higher education by students, staff, and faculty.

In 2005, FBI Director Robert S. Mueller, III developed the National Security Higher Education Committee that "meets regularly to discuss national security matters that intersect with higher education" (FBI National Press Office, 2009, para. 1). This committee, comprised of prominent university leaders, unabashedly promotes a program that "will . . . assist in the development of research, degree programs, course work, internships, opportunities for graduates, and consulting opportunities for faculty relating to national security" (FBI National Press Office, 2005, para. 2). If colleges and universities cooperate with the Committee, this means that the government will have access to, among other things, "intellectual property developed through US university research, sensitive and classified research, researchers' ability to get first-to-market with their ideas . . ., personal and sensitive information . . ., animal rights terrorism . . ., [and] eco rights terrorism" (FBI, 2011, para. 41). In sum, the state agenda is dominating colleges and universities while both blatantly and covertly policing pedagogy. But what does this culture of fear have to do with community colleges, those open-enrollment institutions that seem so far removed from the lofty universities with which the FBI troubles itself?

To be clear, government agents are not regularly knocking on community college doors as they would at research institutions. Rather, policing on the two-year campus is more subtle and bound to the community itself. Policed pedagogy has regularly manifested as mild frustrations to me throughout my years as an adjunct and now tenured professor at various two-year schools. However, after researching academic freedom and repression more fully, I realized that those frustrations were indicative of a larger problem in higher education through which educators are regularly challenged and restricted as experts in their fields.

Consequently, I argue that academic freedom issues are impossible to divorce from those of campus policing. In his commentary on "challenges to academic freedom" in community colleges, Scott Rausch (2005) notes that such obstacles

> come from internal administrative and bureaucratic forces more than outside pressure and public controversy [as they would at larger institutions] . . . . In community colleges as well as elsewhere, academic freedom is threatened not only by intrusive federal policies, oppressive government, and reactionary public forces, but also by intra-faculty conflicts, bureaucratic turf battles, labor tensions, and politically charged atmospheres within the institutions themselves. (para. 7)

In fact, corporations have actually begun designing curriculum at community colleges, thereby negating the role of the faculty member who is forced to relay information that serves corporate interests. For instance, Cisco Systems, Inc. develops courses for Macomb Community College in Michigan (Wilson, 2010, para. 83). Tye, Tye, and Tye (2010) relate this tendency to Michael Apple's concept of "deskilling," noting the "increasing centralization of decision-making at locations farther and farther from the classroom" (p. 35). Although the Tyes write of K–12 education, this happens at the community college level as well, for community and technical colleges are responsible for service to the community, and they are held directly accountable to local stakeholders such as politicians and business leaders who may direct course content (Franke, 2006, para. 4).

To a certain extent, colleges and universities have always been a business, with a hierarchy of power and the inability to survive without patrons. In fact, their reliance on temporary labor actually started out in 1970 as a response to a financial crisis in which nationwide enrollments were expected to be low. However, Schell and Lambert Stock (2001) observe that this temporary change became permanent even when those enrollment issues alleviated (p. 2). Whether one has an M.A. or a Ph.D., as part-time faculty, one is cheap labor and easily disposed of. Consequently, it's no wonder that adjuncts are now the new majority faculty at community colleges. Berry (2005) paints adjuncts as "a group that has experienced proletarianization in nearly all of its classical components: declining wages and job security, loss of ancillary compensation, loss of autonomy and control of work process, and, finally, loss of the (professional) prerequisites that have traditionally gone along with the work of a (tenured) college teacher" (p. 17). As often emerges in the police state, citizens know they are being watched and that they could easily be made to disappear; thus, they desire to keep power holders happy for fear of the unemployment line.

Growing reliance on contingent faculty breeds a culture of fear on college and university campuses, with Nelson noting that course material can become "tame" when part-time instructors fear losing their jobs (quoted in Wilson, 2010, para. 15). This issue should not be overlooked, for Wilson (2010) reports that as of 2007, 69 percent of community college faculty are part time, as opposed to 32 percent at four-year institutions (para. 8). As contingent faculty can face non-renewal if they teach controversial material that rankles students, administrators, other faculty members, parents, or community members, they are bound to teach in a fear-based environment in which administrators can capriciously decide not to rehire them with the simple, and not necessarily factual, excuse that there are no more courses to teach. There are many stories of this happening to adjunct instructors (see Franke, 2006; Wilson, 2010), and contingent workers are faced with the sobering truth that there are *many* other individuals eager to take their positions if they are forced to leave campus.

Wilson (2010) details Nelson's comment that the "most chilling stories [he hears] are from faculty who withhold controversial material from their syllabi . . . . They don't want students or administrators to get upset" (para. 16). Wilson further quotes Mike Van Meter, a contingent faculty member from Central Oregon Community College:

> I recognize some students will hate me if I'm as tough as I need to be . . . . A single student complaint can be very scary for some adjuncts. Part-timers don't know if they will work the next term, and they have to pay medical bills, like everyone else. And what would happen if I were to speak up about a college bond measure? The reality is, it could affect the way the college would view my employment. (para. 18)

The recurring theme here is that faculty are scared, and that is why I have attempted to paint this chapter as a microcosm of the larger issues addressed in the rest of the book, to make connections between policed campuses, policed pedagogy, policed students, and their inherent connection to surveillance of college and university content. All of these issues are interrelated. I further argue that this fear comes from the corporatization of colleges and universities, with teachers being more accountable to the government, local, and national corporations (buttressed by administrators) than to students.

Policed, or what may also be termed fear-based, pedagogy, also manifests for probationary and contingent faculty via course evaluations. Here, the adjunct cannot win, hence the tendency, as Van Meter notes above, to want to be liked by students more than to teach them. If a part-time faculty member gets poor evaluations (i.e., too many negative answers in response to "Would you recommend this instructor to another student?"), there is a likely chance that she will not be rehired. If she receives excellent feedback, it ultimately means nothing to her finding a permanent position at a given institution, especially if she is teaching controversial material, critiquing school policy, or if an administrator simply does not like her. The only payoff to this work, therefore, is the interaction she has with students, but even that is tenuous under current conditions as students buy into the corporate model and begin to see themselves as customers seeking degrees rather than learners seeking educations.

Campus administrators know and take advantage of the adjunct's fear. As reward for her years in graduate school and in response to her desire to educate, the adjunct and the rest of her co-temporary-TBA-flexible-contingent-adjunct-at-will-casual-staff-workers "maintain the fantasy that they really exist elsewhere, in some place other than the overwhelmingly excremental testimony of their experience" (Bousquet, 2003, p. 138). All of this emerges from the culture of fear, and fear is a driving component of the police state in which the power to control the majority resides in the hands of a small minority. This is happening at all levels of higher education, and although the manifestations may differ, the outcomes are similar.

Personal Case Studies of Policed Pedagogy

Nocella, Best, and McLaren (2010) report that if one looks at the history of academic freedom, pedagogy is primary: "The concept of academic freedom has roots in the High Middle Ages when medieval scholastics developed critical pedagogical methods to liberate students from authorities in order to freely pursue the truth" (p. 24). Community college instructors, especially in English departments, have less opportunity to teach high theory and, in essence, to be radical truth-seekers, but we are most certainly interested in fine tuning our "critical pedagogical methods" to enable students to become truth-seekers in the vein of medieval scholastics. Community college instructors teach general education, survey, and pre-college-level developmental courses that leave far less opportunity for teaching one's graduate school specialty. In many cases, we help students build the skills necessary to critically assess their culture and whatever rightist, moderate, or radical messages they will receive later in life.

As much as I would like to teach my pre-college-level students about, say, anarchist principles, they often arrive in my classroom feeling betrayed by a high school system that allowed them to graduate without acquiring basic writing skills that will allow them to fill out a résumé or write an effective cover letter. Before they can understand social rebellion, they must first learn to articulate themselves both verbally and through writing. This is what my students want. While I can and do intersperse subversive ideas into the pedagogical mix, I would be doing my students a disservice if they were to pass one of my courses still unaware of how to form a complete sentence.

In a pre-college-level class during my first year as a community college tenure-track probationary faculty member, I encountered a prime example of what I see as policed pedagogy; in this case, policing manifested as exit exams. Fortunately, my current school of employment has just recently called a moratorium on exit exams, which are final tests used to determine if students may pass their pre-college course and move on to the next level, which may be another developmental class or Freshman Composition, a 1000-level course. Many two-year colleges still use these types of exams, but I believe they exemplify what Nelson (2011) categorizes as "counterproductive assessment and testing protocols [and a] nonconsensual imposition of pedagogical conformity" (p. 38). Exit exams strengthen "the idea that everyone [should] have to reach the same level of proficiency or be punished for falling short," and they ignore issues of "socioeconomic status and family background, including English language proficiency" that should be correlated with test scores (Tye, Tye, and Tye, 2010, p. 32). Nelson further notes that such "impositions" need to be challenged, for by challenging oppression on the micro-level, as I attempt to do in this chapter, I can also challenge it on the macro-level, as evidenced throughout the other chapters in this text.

In the case of my Fundamentals of Writing class, at semester's end, students were expected to write a one-paragraph essay that would then be graded not by me, their instructor, but by others from our English Department. In turn, I would grade exit exams from other classes to determine if another faculty member's students, whose identities remained anonymous, were ready to move to the next level course. These exit exams occurred after all or almost all other course work had been completed. This means that theoretically, a student could have earned an A in my class based upon work done on assignments that I designed, but still fail the class if an outside instructor did not pass their one-paragraph essay. Being new to the college at the time, I questioned if that had ever happened before—a student passing all course work but then failing the exit exam graded by another instructor. I was told that although rare, it does happen.

The unreasonable nature of this policy was not lost on my students. Some were patently hostile to it, and many chose to write their final course paper as an argument against exit exams. Of course, I completely agreed with their views, but I, like they, felt trapped by an unfair administrative procedure. It was my first year as a tenure-track instructor, so I did not want to cause trouble. It was their first year as college students, so they did not want to cause trouble either. In other words, we all fell prey to the culture of fear that arises when curriculum is policed. At the time, I felt that I could only do my best to prepare them for the exam. Fortunately, everyone who passed my course also passed the exam, although this did not suddenly make the policy fair. In fact, it seemed antithetical to my school system's academic freedom policy, which states that instructors have the right "to evaluate student performance"; however, this right is prefaced by the caveat that faculty members must also "follow course outlines as developed by and with colleagues in the department(s)" (MSCF, 2005, p. 113). In this case, one aspect of the course outline was the exit exam.

Regardless of the suspect wording, these paradoxical "rights" restrict individual instructors from evaluating their own students. The exit exams seemed to negate the work we had done and the relationships we had built over the semester, while also posing an overt challenge to my ability to assess those who had enrolled in my Fundamentals of Writing class. As instructor, I was stripped of my ability to make final determinations with my students, and in essence, my pedagogy was surveilled and determined by outside forces.

A similar incident occurred during my next year as a probationary faculty member. In 2010, I was asked to develop a syllabus for a proposed Latino/a literature course. Latina literature and theory partly comprised my doctoral qualifying exams, and I have publications in that field, so I was seen as a prime candidate to develop such a course. My colleagues and I were responding to student requests for more ethnically diverse classes, and we felt that such a course would benefit a diverse student body with a growing Latino population. Recent studies suggest that Minnesota's "Latino population is projected to rise rapidly, growing from an estimated 196,300 in 2005 to 324,400 in 2015 and 551,600 in 2035" (McMurray, 2009, p. 1). These forecasts indicate that learning about Latino culture and history could benefit both those within the growing Latino community and non-Latinos who may be unaware of that culture and history. To this day, the course has not run, and it is not likely to. Unfortunately, my ability to offer this course was obstructed on multiple levels, and I continually felt as if my expertise as an educator and scholar was challenged (a.k.a. policed) for political and financial reasons, not pedagogical ones.

First, the proposal was challenged by the World Languages Department because I wanted to teach Meso-American poetry in translation. This aspect of the course was vital to develop an historical context showing students the rich history that Latinos have in what we now refer to as the South, Central and North Americas. My plan was to establish context by having students read Aztec and Mayan poetry, as well as works by the 17th-century Mexican poet Sor Juana Inés de la Cruz. However, by doing so, I was infringing on another department's "turf," to use Rausch's term. Even though the works would be in translation, World Languages felt that it was their job, not that of the English Department, to teach Latino culture and history. But of course, it is nearly impossible to teach any literature course without including some historical foundation. After an email debate, proposal, and syllabus revisions, and attendance at the approval committee's public meeting, the course was finally approved with the caveat that I tweak my syllabus description to differentiate what I would be teaching from what might be offered by World Languages. Although wearisome, this hurdle was fairly easy to jump as long as I was willing to revise the verbiage. Not wanting to upset administrators or another department, I acquiesced.

This course was then offered online by the administration, effectively ending its ability to run. I was never asked if I wanted a ground or online class, and I was surprised when the schedule came out to see that it was being offered online. As noted above, policed pedagogy determines the "physical place of instruction," which also inevitably alters course content. The rise of online learning exemplifies Apple's (n.d.) "deskilling" process:

> [T]he skills that workers have developed over many years [i.e., critical pedagogy in the classroom] are broken down and reduced to their atomistic units, automated, and redefined by management to enhance profit levels, efficiency, and control. In the process, the employee's control over timing, over defining the most important way to do a task, and over the criteria that establish acceptable performance, are slowly taken over as the prerogatives of management personnel who are usually divorced from the place where the actual labor is carried out . . . . One of the major effects of the current (over) emphasis on technologizing classrooms may be the deskilling and disempowering of a considerable number of teachers. (p. 32)

At its worst, "management personnel" are actually silent observers in the online setting, meaning that they actively monitor individual instructors' online classes. I have directly experienced this when I worked at two for-profit online universities, and although this does not happen at the community college where I work, it is rumored to be on the administrations' agenda.

Online learning is valuable in many ways, but it does not translate well into all types of classes and certainly not for all students. Literature classes are best offered face-to-face. Summaries of and commentary on literary works of all kinds are available on the Internet, and often students can rely on those resources to replace the act of actually reading the literary pieces. I and other English faculty members often speak of how the ability to observe student body language can assist us with assessing how much they understand of a reading, or even if they did the reading at all. If they have not read, I do not seek ways to punish them, but rather to still engage students with literary themes. This skill becomes obsolete in the online setting.

Had I been asked, I would have demanded a ground class for the Latino/a literature course. More significantly, a number of students had reported to both me and another faculty member that they wanted to take the course but decided not to because it was not in a classroom.

However, none of us had a say in how the class was to be delivered. In the end, enrollment was too low to justify running the course. The enrollment cap at my institution is 30 students for an online literature course and 45 for face-to-face literature courses. Without at least half that number, the course will not run, which it did not the first time it was offered, and now there is departmental talk of taking it out of the rotation list.

I am hesitant to make the claim that courses are offered online merely for financial reasons, meaning that the virtual setting allows for more classes to be offered without having to worry about space issues on campus. The verdict is still out as to whether or not online classes actually save colleges and universities money. In a recent *Chronicle of Higher Education* article, Parry (2011) notes university "e-learning" experts such as Joel Hartman of the University of Central Florida have reported that online classes may eventually provide financial gain in the future, but not so in the present. In fact, "generating money is 'one of the worst reasons' to get into online and blended learning" (para. 3). However, enrollment caps are most assuredly determined for financial reasons, much to the detriment of faculty, students, and education in general.

In "The Freedom to Engage," Rhoades (2010) observes that community colleges, much like liberal arts colleges, have traditionally "emphasize[d] their small classes and the personal attention students receive from professors who know their names and are available to engage and mentor them" (p. 55). Unfortunately, this community college selling point is quickly becoming a feature of the past. This is especially problematic for those who teach writing classes. Although the proposal was temporarily shelved, our college president recently announced plans to increase both developmental and college-level writing classes by three students each to help with budgetary issues, an imposition suggested to a faculty body who had already agreed to a pay freeze that is just now ending after three years.

This increase may not sound overly problematic, nor an example of policed pedagogy, but it *is* when considering that most writing classes nationally are already beyond our discipline's recommendations. For example, at my school, both developmental and college-level writing classes are capped at 25 students. However, the National Council of Teachers of English (NCTE) (2008) counsels the following:

> No more than 20 students should be permitted in any writing class. Ideally, classes should be limited to 15. Students cannot learn to write without writing. *In sections larger than 20, teachers cannot possibly give student writing the immediate and individual response necessary for growth and improvement . . . . Remedial or developmental sections should be limited to a maximum of 15 students. It is essential to provide these students extra teaching if they are to acquire the reading and writing skills they need in college . . . .* No English faculty member should teach more than 60 writing students a term: if the students are developmental, the maximum should be 45. (para. 23, emphasis added)

The Association of Departments of English (ADE) (1998) agrees, further noting that "the number of courses taught by each instructor and the number of students in each class, especially in writing courses, has reached unacceptable levels. This problem has become acute in independent and public institutions alike" (para. 2). I have 100 writing students during the current term in which I write this chapter, and I agree with both NCTE and ADE that I cannot possibly give that many students the attention they deserve.

This is policed pedagogy. Administrators devalue education when they put budgetary concerns over pedagogical and student ones, and they are effectively taking control of our classrooms despite expert opinion and faculty narratives that attest to the detriment of such policies. In a typical scenario, if I were to teach three writing courses capped at 28 students along with 45 literature students, this would result in an over 10 percent increase in workload, factoring in the time it takes to give proper attention to student work. In sum, I would also be expected to grade about 470 essays per semester of varying lengths, not factoring in lesson plans, student conferences, and committee responsibilities. As my discipline's organizations argue, such numbers strongly work against my ability to actually help students learn to write, which is the primary goal of community college English departments.

Further, Nelson (2011) argues that faculty should be "skeptical" when institutions claim they are financially strapped, suggesting that "[s]ome administrators exaggerate financial difficulties so that they can reallocate money for unneeded building projects for disciplines they believe can generate income" (p. 39). When my department asked our president to consider other ways to save money, he claimed there were none, but he would listen to any suggestions we had. However, it is unfair to expect English faculty to take control of our school system's financial woes, especially when they go beyond just our college and into our 31-college system. As faculty, we were put into an untenable situation with little recourse but to obey orders from above. Although we "won" the enrollment increase battle because the president was technically breaking union contract by proposing the increases, we have been advised that the enrollment issue will be on the agenda again in the near future, as will talk of standardized writing classes, administrative surveillance of online courses, eradication of tenure in our school system, and other methods of classroom, curriculum, and course domination.

Conclusion

When living in a culture of fear and policed learning, the question becomes: *What do we do about it?* One answer is to involve students in the issues that seem relegated to instructors and administrators. Whether one is a contingent or tenured community college teacher, graduate student instructor, tenured university professor, or scholar, we are all educators. Thus, students are our allies, and they need to be aware of the policing on many levels, especially those I have examined in this essay. In "Same Struggle, Same Fight," Elana Peled and her colleagues (2001) assert that it is only "logical" to make students aware, "for many students arrive at the university believing that their education will serve as their ticket to the American Dream; they are certain that the instructors...have already arrived" (p. 242). In fact, no matter what field students might go into, it is likely that they, as is the adjunct and average American worker, will be "regarded as faceless servants of a flexible fiscal budget, dispensable as the first course of action in an apparent budgetary shortfall" (Peled et al., 2001, p. 234). We are seeing this happen in the workforce today as the gap between rich and poor increases.

When students are taught under a regime of policed pedagogy, higher education becomes a boot camp for the corporate industrial complex, which always involves homogeneous thinking among its workers, and students must be conscious of the ways in which higher education is a microcosm for the world outside its ivy, brick, or virtual walls. Only by seeing the ways in which their instructors and educations are policed may they see the bigger picture of the FBI and its constituents, from campus security to local police departments, gaining control of higher education. Only by seeing this control can faculty and students forge alliances to challenge the system of domination.

Bibliography

Apple, M. W. (n.d.) Computers and the deskilling of teachers. *Rethinking media & technology*. Sindh Education Foundation. http://www.sef.org.pk/old/Educate/education/Computers%20and%20the%20DeSkilling%20of%20Teachers%20by%20Michael%20Apple.pdf

Association of Departments of English. (1998). ADE policy statements: ADE guidelines for class size and workload for college and university teachers of english: A statement of policy. http://www.ade.org/policy/policy_guidelines.htm

Associated Press. (2011, September 20). New MnSCU chancellor promises to shake things up. *CBS Minnesota*. http://minnesota.cbslocal.com/2011/09/20/mnscu-chancellor-presents-priorities/

Barrett, K. (n.d.). Kevin Barrett 9/11 truth cv. http://www.st911.org/truthjihad/cv.htm

Baum, S., Little, K., and Payea, K. (2011). Trends in community college education: Enrollment, prices, student aid, and debt levels. *Trends in Higher Education Series of the College Board and Advocacy Center*. http://advocacy.collegeboard.org/sites/default/files/11b_3741_CC_Trends_Brief_WEB_110620.pdf

Berry, J. (2005). *Reclaiming theivory tower: Organizing adjuncts to change higher education*. New York: Monthly Review Press.

Bousquet, M. (2003). The waste product of graduate education: Toward a dictatorship of the flexible. *Works and Days: Information University: Rise of the Education Management Organization, 21*(1–2), 129–152.

FBI. (2011). Higher education and national security: The targeting of sensitive, proprietary, and classified information on campuses of higher education. http://www.fbi.gov/about-us/investigate/counterintelligence/higher-education-and-national-security

FBI National Press Office. (2005, Sept. 3). FBI director appoints national security higher educational advisory board. http://www.fbi.gov/news/pressrel/press-releases/fbi-appoints-national-security-higher-education-advisory-board

——. (2009, Nov. 3). National security higher education advisory board meets at FBI headquarters. http://www.fbi.gov/news/pressrel/press-releases/national-security-higher-education-advisory-board-meets-at-fbi-headquarters

Franke, H. (2006, October 27). How strong is academic freedom in community colleges? *Chronicle of Higher Education*. http://chronicle.com/article/How-Strong-Is-Academic-Freedom/13551/

Giroux, H. (2010). Higher education after September 11th: The crisis of academic freedom and democracy. In A. Nocella, S. Best, and P. McLaren (Eds.), *Academic repression: Reflections from the academic-industrial complex* (pp. 92–111). Oakland, CA: AK Press.

Kunstler, W. (2004). Public ethics and the Bill of Rights. In W. Kunstler, M. Smith, K. Goldman, and S. Kunstler (Eds.), *The Emerging Police State: Resisting Illegitimate Authority* (pp. 15–28). New York: Ocean Press.

McMurray, M. (2009). Minnesota population projections by race and hispanic origin, 2005 to 2035. Minnesota State Demographic Center, 1–7. http://www.demography.state.mn.us/documents/MinnesotaPopulationProjectionsbyRaceandHispanicOrigin2005to2035.pdf

MSCF. (2005). Master agreement between Minnesota state colleges and universities, board of trustees and Minnesota state college faculty. http://www.hr.mnscu.edu/contract_plans/documents/mscf0507.pdf

National Council for Teachers of English. (2008). More than a number: Why class size matters. *NCTE Guidelines*. http://www.ncte.org/positions/statements/whyclasssize matters

Nelson, C. (2011). A faculty agenda for hard times. *Academe*, 97(3), 38–40.

Nocella, A. (2006). Fighting against the conservative agenda in the academy: An examination of the 4Ss of academic repression and repressive pedagogy post–9/11/01. *Journal for Critical Education Policies*, 4(2). http://www.jceps.com/ index.php?pageID=article&articleID=73

Nocella, A., Best, S., and McLaren, P. (2010). Introduction: The rise of the academic-industrial complex and the crisis in free speech. In A. Nocella, S. Best, and P. McLaren (Eds.), *Academic repression: Reflections from the academic-industrial complex* (pp. 13–89). Oakland, CA: AK Press.

Parry, M. (2011, March 30). Think you'll make big bucks in online ed? Not so fast, experts say. *Chronicle of Higher Education*. http://chronicle.com/blogs/wiredcampus/think-youll-make-big-bucks-in-online-ed-not-so-fast-experts-say/30663

Peled, E., Stock, P., and NCTE. (2001). Same struggle, same fight: A case study of university students and faculty united in labor activism. In E. Schell and P. Lambert Stock (Eds.), *Moving a mountain: Transforming the role of contingent faculty in composition studies and higher education* (pp. 233–244). Urbana, IL: NCTE.

Rausch, S. (2005). Challenges to academic freedom in community college history programs. *Forum on Academic Freedom*. http://www.oah.org/pubs /nl/2005may/rausch.html

Rhoades, G. (2010). The freedom to engage. *Academe*, 96(3), 55.

Schell, E., and Lambert Stock, P. (2001). Working contingent faculty in[to] higher education. In E. Schell and P. Lambert Stock (Eds.), *Moving a mountain: Transforming the role of contingent faculty in composition studies and higher education* (pp. 1–44). Urbana, IL: NCTE.

Tye, D., Tye K., and Tye B. (2010). The slow death of the American teacher. *Obama and Education*, 5(1). http://opus.ipfw.edu/cgi/viewcontent.cgi?article=1003&context=spe

Wilson, D. (2010). The casualties of the twenty-first-century community college. *Academe*, 96(3), 12–18. Retrieved from Academic Search Premiere Database.

• CHAPTER FIVE •

Of Accountablity, Surveillance, and Fear

Speaking Out and Losing My Job

Barbara Madeloni

I am a teacher educator at a public university who is both subject to and a participant in the system of "accountability as surveillance" currently encroaching into higher education. Teacher education, with its close ties to K–12 education, low status in the academy, and relationship with state mandates, is easy pickings for the profiteers who use accountability and surveillance as tools to commodify teaching and learning, and to privatize public education. As a critical pedagogue, I base my practice on a commitment to education as liberation, and to developing new teachers with political and social consciousness, teachers who know how to use their voices and make space for the voices of their students to not only speak, but to also act to create a just world. As such, I am committed to not only teaching about social justice education, but enacting social justice practices in my classroom and my community. For doing just that, for making a space for the student teachers with whom I work to identify and reject corporate encroachment into teacher education, I received a letter of non-renewal: I lost my job.

In the spring of 2012, I facilitated discussions with pre-service teachers in the student teaching seminar about the demand by the university that they participate in a field test being run by Pearson Inc. through a contract with Stanford University of their Teacher Performance Assessment (TPA). The student teachers, alert to the dangers of incursions by corporations into the public sphere, aware of the risks to privacy when we give information to private corporations, and mistrustful of an assessment tool that they experienced as inauthentic being scored by out-sourced labor on a piece-work contract, demanded the opportunity to opt in or out of the field test. Citing ethical standards for human subjects reviews, students demanded they be given informed consent about the pilot study, its purposes, and the security and confidentiality of their work. For three months, university officials obfuscated, insisting first that informed consent was not required as it was not a "research study" and then asserting that consent came with uploading work to the data management system—work that was required in order to complete the program. Undeterred by this attempt at coercion, students continued to insist that they did not want to hand over their intellectual property or videos of themselves and their students to a for-profit corporation. Local school officials, teachers, and parents joined these students in objecting to the infiltration of a private corporation into the work of developing and assessing new teachers. When university officials finally conceded that the students were correct in their demands, 67 of 68 secondary education student teachers chose to opt out of participation with the Pearson field test.

Shortly after the students' victory in standing up to Pearson and the university, Education Radio (2012) produced a program about their resistance. One outcome of this program was that student teachers at other universities contacted the radio show to say that they too felt coerced to participate in the field test. Energized by the students' success, anxious to share their story and inspire others, and certain that there were more voices to be heard speaking back to Pearson, I contacted a sympathetic reporter at *The New York Times* (Winerip, 2012) and invited him to do a story about the resistance. He did. Two days after the story appeared, the associate dean and the department chair demanded a meeting. In this meeting I was told that I was "to be scrupulous" about the power I held over students and to always be very clear that I do not speak for the university. At the same time, work that had previously been assigned elsewhere was assigned to me. Seventeen days after the article appeared in the *New York Times*, I received my letter of non-renewal.

How Did We Get Here?

The Discourse

Lines of police officers in riot gear and using pepper spray do not appear over night. They emerge within increasing surveillance, fear, and the discourses that accompany these. In education, the narrative of accountability has been central to setting the stage for more policing of the work of academics and K–12 teachers. Apple (2005) has been warning us for some time of the dangers of audit culture in education. What he refers to as the "odd combination of marketization on the one hand and centralization of control on the other" (p. 11) manifests in increasing focus on efficiency, production, and measureable outcomes. This focus, which imposes a business model on all interactions, requires increasingly high levels of surveillance of its worker subjects to assess their efficiency and output. Apple writes of how the discourses of audit culture—data driven, standardization, measurement—infiltrate our thinking and become common sense. But the centralization is as critical as the market ideology in audit culture, because it is through centralization that surveillance and its cousin, fear, enter our work.

Writing from inside this audit culture, I am well aware of how fear implicates itself into our lives through the discourse of accountability. It comes first in the form of colonization of our time. Audit culture is nothing if not profligate in its ability to create tasks, rubrics, data points, data reports, etc. Each comes with its own set of technological tools, available for a fee, which must not only be paid for, but learned, taught, practiced. At the micro-level of actually using these devices in the classroom, the tasks themselves become the masters of the classroom experience, not the students, not the instructor, not the community of the classroom. If we take the Teacher Performance Assessment as an example, the 40–46 pages of detailed assignments not only lead students to answer only the specific questions determined by external forces to be relevant to their student teaching experience, but narrow the possible "good" answers through the limitations of a rubric. In a high-stakes environment, where the score on this "assessment" determines a student's ability to graduate and/or receive a teaching license, the production of the right answer dominates the experience. I am astonished at the ease with which participants in these technocratic exercises insist that the lived experience is not changed by the imposition of these formulas. But my astonishment diminishes when I understand that the common sense discourses of education have come to deny the human beings in all of their sloppy beauty as actors in the experience. Our humanity gets in the way. It troubles the data. It holds within it possibilities of freedom and rebellion. If we eject the human from what and how we "measure," we eliminate the opportunity to note that something is amiss. The contortions I am invited to do to make room for these dominating audit tasks deny the degree to which they change me, the students, our understanding of teaching and learning, and the ways we come to know each other in this work. This initial move toward responding to the external mandate rather than the shared lived experience is the first stage of the development of a police state within education.

This policing is not simply a metaphor, but with the availability of new surveillance technologies, a fact. The TPA requires student teachers to send videos of their classrooms to Pearson to be viewed and scored by contract labor. Within my university, surveillance methods are used to track faculty use of the assessments, reviewing all TPAs after faculty have reviewed them. The data management system that we use can be accessed for reports of when I log on and for how long. Student work, within the Pearson TPA system, is viewed and scored by unknown scorers who have been "calibrated" to respond appropriately to the demands of the rubric. All of this serves to normalize the invasion of our privacy, while generating just enough anxiety to elicit our conformity. This level of oversight is bound to impact how we know ourselves as well as our students. It brings with it the threat of admonishment, at the very least, for wrongdoing.

This shift, wherein we no longer respond to the people with whom we teach, learn, make meaning, but instead to centralized mechanisms of control, is, as Biesta (2004) notes, contrary to actual responsibility. The centralization of authority removes us from the human conditions and relationships of responsibility. Thus, we get Apple's new managerial class, whose skills in processing data and efficiency measures allow them mobility and power within the neo-liberal state. The discourse of accountability further offers these new managers a narrative of moral righteousness as they use the power of the audit as "neutral instrumentalities" to solve the problems of the world through accountability and efficiency. It is essential, as we come to understand the growth of the police state, that we attend to these places where our minds become occupied with ideas, practices, words, and values that are contrary to how we know ourselves as human beings in community and in democracy.

The abstraction of the panopticon is a lived experience. When I spoke out in public forums and questioned the TPA as an assessment, I was reprimanded on my annual faculty report and told I needed to *work with* the department chair on the TPA (her emphasis). The implication, which I challenged, was that my critiques, rather than being an expected part of the intellectual culture of the academy, needed to stop. When I was admonished as if I had exercised too much power over the students who objected to the Pearson TPA and told to never speak for the university, the deeper process message was to both tarnish and isolate me. The impact of a dominating discourse cannot be exaggerated, especially when it becomes attached to material and social reward systems. As a contract faculty, I was aware of the risks I took when I challenged the implications of the tasks I was told to complete for teaching, learning, and public education. Indeed, my job vulnerability and that of other contract faculty is precisely why we were given these tasks. The expectation was that fear would silence us. However, it is important to note, as a teacher educator in a licensing program, I was both surveilled and surveillor. My job involves reviewing assessment data, assuring that others complete various protocols and audits. I felt myself as implicated, and the shame of this self-awareness invited me to silence myself, to bury my head and just go on within the system. Self-policing and policing within the "professional" community is an essential step in preparing us to be policed more forcefully.

A critical element in the exercise of control is through shaping what we do and do not attend to. Standards and market-model accountability systems draw our attention both to and from. In a world in which we only attend to explicit data points to measure and know, the implicit meanings that we understand within processes, within lived experiences, are denied. Some years ago I received a poor evaluation of my work as a high school English teacher because I focused too much on process. As time passes, and from living within systems whose processes are dehumanizing, I understand the danger of attending to process. Attention to process is attention to what is actually happening to us. Thus, the Teacher Performance Assessment does something to us: it both communicates and demands compliance to a rigid, externally imposed perspective on ourselves, our work, and our judgments of both. More, within the institution where I work, the day-to-day process I experience is one in which to be "professional" is to respond obediently to mandates, standards, and auditing. We experience these as people, as feeling, social, complex human beings. We make meanings within these experiences about who we are and who we can be. We are shaped. And while some might tell us that we also shape, it is imperative that we attend to issues of material power. I responded to the shaping of teacher education the TPA imposed by reshaping the conversation. I lost my job—a real material result of the policing of the academy.

Why did I take this stand about this issue? Colleagues and administrators told me that the TPA is inevitable, that we should learn to work within it, and that it is serving its function to improve teacher education by being rigorous and objective. I had a colleague once suggest that I should not lead students to think they were in prison—an interesting metaphor. I wondered with her, what if we are? The policing of our minds, of our expectations for ourselves, for our students, for our practices, begins with these "performance" demands. These demands carry with them ideas of knowledge, of purpose, of how we live our lives, which are contrary to democracy. Further, in their dehumanizing processes and ready acceptance of surveillance, they open the door for the police state. Apple asks us to consider what we defend when we defend the academy. I say, we defend the places where we meet in a relationship of human beings—one to the other—in all of the multiplicities of our knowledge and experience. We defend that archaic possibility: our humanity.

Power and Profit Motive

What are we defending against? Audit culture as we experience it does not exist apart from our capitalist systems and structures. Its reliance on surveillance and policing the individual serves particular interests. These interests are those of the corporate capitalists whose motives are greed and profit. That we are under siege from the assault on democracy and the commons by privatizing forces is well documented. (Watkins, 2011; Carr and Porfilio, 2011). This assault is many pronged and has been successful, in part, because of its ability to control education, broadly speaking (Giroux, 2012), and its ability to access and exploit certain "common-sense" explanations for our experiences (Lipman, 2011). Privatizing forces have disguised their intents by clever manipulations of language, and by infiltrating and compromising leadership in institutions whose mission once was to serve the public good (Newfield, 2008).

In the case of the Teacher Performance Assessment, we find a perfect example of the intersection of a for-profit company (Pearson, Inc.), the university (Stanford and the University of Massachusetts, as well as other colleges and universities across the country), professional organizations (the American Association of Colleges of Teacher Education) and the state (Massachusetts Department of Elementary and Secondary Education and other state departments of education nationwide). Pearson Inc. is a multi-national and the largest assessment company in the United States. On its website, Pearson notes that in the past five years its profits from educational materials and services has almost doubled and that its long-term strategy includes investing in educational content and digital markets. "We aim to be the world's leading learning company, serving the citizens of a brain-based economy wherever and whenever they are learning. In financial terms, we measure our progress against three key measures: earnings, cash and return on invested capital" (Pearson, 2011). Their reach into education begins in pre–K and extends through higher education with their hands in testing, the curriculum for the new common core standards, school improvement turnaround plans and contractors, the General Education Diploma, online courses, and myriad other educational "products." Indeed, Pearson have been the leader in exploiting the new education markets touted by Joanne Weiss (2011), Secretary of Education Arne Duncan's chief of staff, when she writes, "The development of common standards and shared assessments radically alters the market for innovation in curriculum development, professional development, and formative assessments. Previously, these markets operated on a state-by-state basis, and often on a district-by-district basis. But the adoption of common standards and shared assessments means that education entrepreneurs will enjoy national markets . . ."

Privatizing for profit and the ability to scale up to access national markets are, as Weiss unashamedly noted, the core value behind so called "education reform." As capitalism is about profit, this comes as no surprise. What is surprising is the way universities, the state, and professional organizations are collaborating with these profiteers. The most generous explanation is that those who lead these institutions are subject to the same dominant narrative as the rest of us. Thus, their acceptance of the terms of the market as the natural context for our lives and choices is no worse than anyone's acquiescence to that narrative. Couple this with the draconian cuts to public universities, and the logic of capturing resources in order to protect the institution takes on a certain unexamined common sense. Returning to Apple's discussion of the new managerial class, we can understand that administrators see the potential for mobility and access within an increasingly inequitable economic system when they agree to the practices of audit culture. At a cultural moment when the individual is left alone to manage the exigencies of life in a cruel and unforgiving environment, the choice to think only of one's personal well-being is to be expected. Indeed, leadership in professional organizations, the university, and state agencies, as well as research agendas are all subject to the commodification as is everything else in what was once the commons. Granting agencies, more and more overseen by corporatists and venture philanthropists, control the language of our research questions as well as the ways we collect information and assign meaning to our experiences. But does the ubiquity of the free-market mentality offer both an excuse and an explanation for the complicity of leadership in higher education with its undoing?

As of this writing, I have no information about the nature of the contract between Stanford and Pearson Inc. regarding the distribution and scoring of the Teacher Performance Assessment. Nor do I have any information about any contractual obligations between Pearson and state agencies that have accepted the TPA as their measure of student teacher readiness. I do not have any information about current or potential benefits to individual faculty members whose work with the TPA might lead to opportunities within and beyond the academy. I do know that there is concern over the link between Pearson curriculum development for the common core standards that are being adopted as part of Race to the Top by many states, the requirement in the TPA for students to show evidence of teaching to these same standards within the TPA, and the expected new revenue for Pearson from the testing that will result from implementation of common core standards. I stand aghast in the face of these connections. How do we come to accept the propriety of this money grab? How do we not understand how the considerations of capital undermine education for democracy?

Part of the effectiveness of the police state within the academy is the willingness of academics to deny the realities of power and the relationships between power and money. Time and again I hear the refrain from the administrators and some faculty colleagues at the School of Education, a refrain we hear from complicit union leaders, faculty, and administrators nationwide, that we need a "place at the table" in policy discussions. In the Commonwealth of Massachusetts, for the discussions of new requirements for teacher education programs, those at the table included the following national corporate education reform organizations: National Council on Teacher Quality (NCTQ), Education Sector, Education Trust, the Data Quality Campaign (DQC), and the Center for American Progress (Massachusetts Department of Elementary and Secondary Education). To pretend that a place at the table stocked with representatives of mammoth corporate interests is anything but a pretense, that it will not result in your becoming the meal, is to exist in a world of profound denial about the material realities of power and voice. It is to deny the actual processes at work in which money buys access, in which power is ceded to those who comply, in which discourses are dominated by those with influence, in which the mechanisms of control are embedded in policies that determine practice that isolate and eventually expel other ways of thinking, being, imagining. University leadership that consists of those who have embraced the corporate ideal or those who are hedging their bets and riding the free-market wave are successful because they deny and invite us to deny how power works.

Standing Up, Speaking Back, Losing My Job

From the first time that the Teacher Performance Assessment (TPA) was introduced as a mini-pilot to the program where I teach, I was cautious. An initial group of three students was paid to participate in the mini-pilot. They were adamant that the assessment, while interesting as a way to talk about their teaching, should never be used as an objective assessment. The following year, when I was told that I was to attend a two-day workshop in which I would be "calibrated" to score the TPA, my alarm grew. It is important to understand that this calibration was a standardization of my work, of the work the doctoral student supervisors, and that of the cooperating teachers—a standardization that denied our experience, knowledge, values, and relationships with students. In a paper exploring the impact of standards in teacher education, Delandshere and Arens (2001) note that those who are most accepting of standards are operating without a philosophy of education, while those of us with a stated philosophy of education are most likely to object to standards, to see what is lost. As an educator with a profound commitment to education for democracy, to education as liberation, to education as what emerges within the context of community, relationships, and dialogue, I experience the TPA as an incursion into the core of my practice.

I spoke out whenever the opportunity arose or whenever I could create the opportunity, though I was never invited to decision-making meetings. I raised all of the questions that I have raised in this chapter. When I learned that Pearson had signed a contract with Stanford to distribute and score the TPA, first during a nationwide field test and then with plans to oversee the entire system in whatever states adopted the TPA, a possibility emerged to put a stick in the spokes of the juggernaut.

I entered this struggle with full knowledge that it would upset the people in power. Throughout the time that I discussed the issue with Mike Winerip, the reporter from *The New York Times*, he asked me numerous times if I was sure I wanted to go forward with the article given the attendant risks. After Education Radio ran the first program about the student resistance to the TPA, Pearson was Googling my name within 24 hours. My choice to speak out came down to this: if I find myself thinking that I live in a world where it is too dangerous to speak out, that is when I absolutely must speak out.

The police state, as I experienced it on my campus, comes in the form of data collection and assessments supported by professional organizations, overseen by university faculty and state functionaries, and owned by corporations. It has insinuated itself through language and surveillance to leave little to no room for my voice, for my experience, knowledge, and humanity or that of the students. People have asked, "Why the TPA, Barbara, when there are other issues?" My answer: because it was here that I could feel myself being erased. The letter of non-renewal proves how desperate the people in power are to silence critique, to demand conformity through fear. The experience of solidarity that the students and I shared teaches me of the strength of relationships and dialogue against the dehumanizing corporate state. In response to my speaking out, I have received numerous letters and emails of support. Organizing is underway to demand I be given my job back and the university address the larger context of the assault on public education. This inspires and encourages me. I have a deeper appreciation for why we must refuse their language and narratives, organize, and revolt through a deep attention to our shared humanity. That is what frightens them. That and the courage it brings us.

Bibliography

Apple, M. (2005). Education, markets and an audit culture. *Critical Quarterly*, 47(1–2), 11–29.

Biesta, G. (2004). Education, accountability and the ethical demand: Can the democratic potential of accountability be regained? *Educational Theory*, 54(3), 233–250.

Carr, P., and Porfilio, B. (Eds). (2011). *The Phenomenon of Obama and the agenda for education: Can hope audaciously trump Neoliberalism?* Charlotte, N.C.: Information Age Press.

Delandshere, G., and Arens, S. (2001). Representations of teaching and standards-based reform: Are we closing the debate about teacher education? *Teaching and Teacher Education, 17*, 247–266.

Education Radio. (2012, March 25). Teacher performance assessment: A money grab for Pearson. http://education-radio.blogspot.com/2012/03/teacher-performance-assessment-money.html

Giroux, H. (2012, June 19). Beyond the politics of the big lie: The education deficit and the new authoritarianism. *Truthout.org*. http://truth-out.org/opinion/item/9865-beyond-the-politics-of-the-big-lie-the-education-deficit-and-the-new-authoritarianism

Lipman, P. (2011). Neoliberal urbanism, race and urban school reform. In W. Watkins (Ed.), *The Assault on Public Education: Confronting the Politics of Corporate School Reform.* New York

& London: Teachers College Press.

Massachusetts Department of Elementary and Secondary Education. (2012, April 19). Proposed amendments to 603 CMR 7.00–Regulations for educator licensure and preparation program approval. http://www.doe.mass.edu/boe/docs/2012-04/item5.html. http://www.doe.mass.edu/boe/docs/2012-04/item5.html

Newfield, C. (2008). *Unmaking the public university*. Cambridge, MA & London: Harvard University Press.

Pearson Inc. (2012). Financial information: Five year record. http://www.pearson.com/investors/financial-information/financial-highlights/five-year-record/

Pearson Inc. (2011). Investors strategy. http://www.pearson.com/investors/strategy/

Watkins, W. (Ed.). (2011). *The assault on public education: Confronting the politics of corporate school reform*. New York & London: Teachers College Press.

Weiss, J. (2011, March 31). The innovation mismatch: "Smart capital" and education innovation. *Harvard Business Review Blog Network*. http://blogs.hbr.org/innovations-in-education/2011/03/the-innovation-mismatch-smart.html

Winerip, M. (2012, May 7). Move to outsource teacher licensing draws protest. *New York Times*, p. 14.

PART II

The Surveilled Campus

• CHAPTER SIX •

Cameras and ID Card Swipes

Disappearing Privacy and the Cultivation of the Virtual Self

Richard Van Heertum

Since the publication of Foucault's *Discipline and Punish* (1995), much has been written of the panopticon and its relationship to schooling and knowledge production. Many have argued that the very organization of schools delimit the space for students to be creative and to think independently. Since the Columbine tragedy, and particularly since 9/11, increased security and surveillance on high school and college campuses has heightened the sense that students are considered dangerous and must be controlled. The 2007 Virginia Tech massacre, where 32 were killed and another 25 injured, only amplified these concerns. New technology has facilitated the response—a world where students are under constant surveillance and thus arguably self-regulate their behavior to meet the dictates of the new social order.

The university was once seen as the space where one could break free from the constraints of one's parents and the rather rigid K–12 system. As bell hooks once argued, "the classroom remains the most radical space of possibility in the academy" (1994, p. 12). Yet have new forms of surveillance and discipline undermined the intellectual and social freedom once proffered in higher education? Are the actual and symbolic powers of discipline and control leading to increased self-regulation and obedience in students? And further, has the advent and increased usage of new media and technology constricted the freedom once associated with college life? One can think of blocked Internet sites in libraries, restrictions on campus networks (students at UCLA were famously sued a few years ago for copyright infringement for downloading songs), and the various ways that outsiders (including the community) face barriers in accessing the knowledge and capacities of college campuses.

In this chapter, I will discuss the use of surveillance cameras and ID card swipes as symbolic of the "surveillance society" that reigns on many college campuses today. I will argue that the concept of the panopticon continues to be relevant, but that we must also consider arguments of the post-panopticon and synopticon as supplementing the original Foucault formulation. Essentially, I argue that new and traditional modalities of surveillance not only lend themselves toward internalization of power by students and faculty, but fundamentally alter the nature of the university and knowledge production, supporting and reinforcing the central tenets of neoliberalism—including privatization, deregulation, the withering of the state and the rearticulation of citizenship as work and consumption (Bourdieu, 1998; Giroux, 2004; Torres and Van Heertum, 2011). Considering in particular the profound influence of new media and technology, I argue that we are creating new subjectivities and dispositifs that challenge traditional notions of privacy and the individual subject, creating more insidious forms of discipline and control.

The Rhetoric of Campus Surveillance

Foucault's *Discipline and Punish* (1995) contemplated the power of architecture and institutional norms to alter the modes of control in society and much subsequent research turned specifically to schools as a key site of the transformation (Hoy, 1986; Lewis, 2006; Shapiro, 2003; Unwin, 2009). The structure of schools and classrooms, the relationship between teachers and students, security measures like cameras and security guards, and data collection and mining all hearken to the original Foucauldian notion that new modes of discipline and control were altering the nature of citizenship and identity, by using discipline and visibility to create new habits of self-regulation.

Crises like those at Columbine in 1999, Minnesota in 2005, and Virginia Tech in 2007 have played a key role in tipping the balance of the dialectic of security and freedom toward security. The social sciences have also played a role in the firming of panoptical logic, by defining the normal and deviant and thus setting parameters that can be monitored and methods to do that monitoring. The effects of both material and virtual surveillance architecture thus serve as intensifiers of power that foster the socialization and normalization of the student population.

A particularly relevant example comes from Biloxi, Mississippi, where post-911 they installed over 500 video surveillance cameras in the schools across their district. While it certainly provided increased comfort and security to parents, the state's American Federation of Teachers worried about issues like the technology adversely affecting teacher rapport with students, privacy, suppression of academic creativity and spontaneity, and the inability of parents, teachers, and students to view the recordings without a court order (Piro, 2008). Similar arguments can be offered of the climate on many college campuses today.

Surveillance arguably sorts people into categories, assigning worth and risk, in ways that have tangible effects on their opportunities and success. Because deep discrimination occurs within the new framework of discipline and control, one could argue that the question of surveillance moves beyond one of privacy to also encompass social justice (Lyon, 2003). Yet I believe that new media and technology have initiated a shift from the panopticon to the synopticon, where students and professors on colleges are not only being watched, but are actively watching each other, creating an environment where all learning and campus activity is structured around a regime of discipline and control. As Btifhaj Ajana argues, we are entering a post-panopticon world while many of the themes of the panopticon continue to persist (Ajana, 2005). Both are modalities of biopolitics, concerned less with the practice of power over the individual disciplined to conform to regulation and more with massification, with the population in its multiplicity controlled and regulated in a preventive way based on modes of intervention that operate at the level of relations between humans (Foucault, 1995). Deleuze thus argues that we have moved from disciplinary society to control society, from centralized power of institutions toward rhizomatic networks of control that proceed far beyond explicit disciplinary deployment of power to more dynamic and implicit forms inscribed in the practices of the quotidian (Hardt and Negri, 2000).

The concept of biopower is quite useful here, defined by capacities and identified by material instantiations rather than subjectivities. It is based on the concept of self-regulation by individuals and communities of their actions, based on particular dispositifs (Rose, 1999). Yet as Ajana argues, we must contemplate the continued relevance of the panopticon together with the post-panopticon arguments, which are based on the ideas of displacement of the panoptical ideal by mechanisms of seduction; redundancy of the panoptical impulse wrought by the normalization of self-surveillance functions; the effects of the socially constructed Western subject on socialization; reduction in the need for panoptical surveillance based on simulation, prediction and action; supplementation of the panopticon by the synopticon; and failure of panoptical control to produce reliably docile subjects (Boyne, 2000).

As Rose (1999) reminds us, however, panopticism was not predicated on producing "terrorized slaves" in a prison-society; instead it sought to construct self-managing individuals formed through the internalizing of power and partitioning of responsibility between state and citizens. Essentially, the argument for the post-panopticon tends to misread the nature of the dynamic and underestimate the power of social formations to be internalized in the actions of subjects, even as they are strengthened by the power of new media and technology to infiltrate the construction of our social and political identities. The new paradigm does incorporate some social changes, though, including mobility over fixity, anonymity over individualization (particularly online), hybridization of techniques, and virtualization of individuality where securitization of identity can appear as a solution.

In this new framework, it is important to recognize that surveillance exists at many levels from the more obvious forms of security cameras and ID swipe technology to the less obvious like the monitoring of online activity and tracking of campus movements. Together the surveillance regime on many college campuses, backed by the synoptical surveillance of everyone by everyone else, creates an environment that ascribes the contours of campus life and the borders that separate members of the college community from within and without. As Ajana argues, surveillance is "the enduring of exclusion for some and the performance of inclusion for others to the point where it becomes almost impossible to demonstrate one's inclusion without having to go through the labyrinth of security controls and identity validation." Balibar further argues that the polysemic nature of all borders creates different phenomenological experiences for different people, creating a "world apartheid" (Balibar, 2002).

Yet how does this relate to campuses? I believe in a number of ways, particularly in creating not only new subjectivities that support the current order of things but by marking knowledge as not for all, but for a select few to use as entry to economic, political, and social spaces that confer status and power. The new surveillance state, in fact, reinforces the notion of education as a separate social space, not part of the real world, but providing entry to it. This notion of the university as a border that must be crossed on the road to success thus serves a role in legitimating social inequality, as the gatekeepers are perceived as fair, objective, and unbiased judges of not only academic potential but really personal worth. By closing off the university and its facilities (particularly the symbolic space of the library, or vessel of knowledge) to those who don't hold ID cards, the university reinforces the notion of knowledge as a private not public good that serves not the needs of society as a whole, but as a more instrumental commodity that functions to confer cultural and social capital and serve the economic needs of society.

Essentially, I am arguing here that the surveillance society thus serves to inscribe neoliberal ideology by indirectly fortifying its central arguments. In this sense, the university becomes a key border legitimating the sorting of individuals within society, often along racial lines. Mary Schmelzer (1993) argues that the entire educational system is intended to ingrain into both students and faculty that normalization is a key part of socialization. Thus self-regulation (fitting in) and individuation is the primary goal of the process, and picking majors, specializations, and being creative are really part of the process of conforming to the larger social norms. John Knight and his co-authors go further to argue that "any adequate analysis of the production, reproduction, and contestation of the social relations of author and power should attempt to also 'excavate' their epistemic foundations, that is those largely unseen controls on discourse whose effects, with that specific epoch, is for permitting certain facts, opinions and ideas to be uttered while forbidding others" (Knight, Smith, and Sachs, 1990, p. 136).

Another way to look at this relationship is as "co-active control," implying individuals who choose to live with ever empowering surveillance technologies that create an alternative types of social control taking over from modern proactive control (Kim, 2004). The classical perspective sees social control as shifting from overt, external, and punitive to covert, internal, and preventative. Yet advanced technologies with effective data-management systems creates more obscure, less intensive, and more powerful means of surveillance, lending themselves toward "pseudo self-control," where invasion of privacy through data capture undermines autonomy and self-regulatory modes. The new paradigm normalizes the notion of proscribed accounts of privacy, citizenship, and free access to knowledge. Facebook, Myspace, YouTube, Twitter, monitored campus discussion boards, and other forms of new technology move surveillance from external agents to individuals themselves, monitoring each other without clear intent to discipline or police the activities they witness.

In this shift to *dataveillance* (Clarke, 1994), data is collected not on but about individuals, a move from the visual to the linguistic, unnoticed watching to intervention, territoriality to structures, centralization to decentralization, and a repressive state to mathematical or instrumental formalism (Kim, 2004). The enhancement of surveillance capabilities then initiates the rise of a highly transparent society where everyone can watch everyone else; what Lyon (1994) phrases "surveillance in everyday life." In a surveillance society, it is our *digital persona* or *virtual self* that matters, a public personality based on data and transactions that serve as a proxy of the individual. Clarke argues this leads to threats to social and individual liberty and autonomy as "each observer is likely to gather a different set of data about each individual, and hence to have a different gestalt impression of that person. Frequently, individuals who are aware of the use of data may be in better position to project data selectively in order to influence the imposed digital persona that is formed" (Clarke, 1994, p. 78).

Institutionalization of information makes it private property and a source of profit and adds to the commodified ways people see themselves and others. This further solidifies the notion of education as just another commodity—a form of knowledge acquisition that provides social and cultural capital. Personal growth external to the ways in which it can be captured in the virtual self are thus peripheral to human existence. Like the precocious child who engages in extracurricular activities simply to ensure entry into an Ivy League school, our activities are increasingly externalized as assembled signs that signify us as collected data snapshots. This whole process loosens the boundary between public and private, increasing the reversal that Bauman discusses where our personal lives become public and public (or political) life is privatized (Bauman, 1999).

As part of the surveillance society, a "snitch culture" has developed, where surveillance is far more insidious and pervasive than anything Orwell ever imagined—with the open access to data about others lending itself toward a world of pseudo-investigative reporting of the quotidian (Redden, 2000). Like the film *The Truman Show*, we come to recognize that our entire lives are being watched, but rather than sail toward the edge of the set to break free, many seem to actively embrace being watched and alter their behavior to fit the contours of the new reality.

Transitioning from the Panopticon to Synopticon

One perfect example of the changing nature of surveillance on university campuses today is the "Dirty 30" list controversy that emerged at UCLA in 2006. A right-wing alumni group called the Bruin Alumni Association offered to pay students $100 for tapes of lectures that showed "radical" professors in practice, backed by a well-researched website that highlighted the 30 most radical professors on campus. Not surprisingly, all of these professors were leftists and noted for not only their stances against conservatives, George Bush, the war, and other conservative positions, but even for supporting Democrats (Wiener, 2006).

The group was led by a single person, Andrew Jones, a 2003 graduate of UCLA who had headed the campus Bruin Republicans, engaging in such pranks as running an affirmative-action bake sale where whites and men paid more for cookies than minorities and women. He also worked for David Horowitz for a short time, before he was fired for allegedly trying to pressure students into reporting that they were physically assaulted by leftists. While Jones's actions were ultimately condemned, the notion that students are watching professors and that professors' futures increasingly rely on positive student evaluation and not "rocking the boat" augurs a world where students reign in the fulcrum of academic power. This is, of course, further cemented by the new notion of students as consumers or clients, whose interests, needs, and desires define the heart of the higher education experience.

Yet this is but one example of an increasing trend toward monitoring, suppressing, and attacking professors who are not complying with the conservative, neoliberal turn on college campuses, reminding us of the McCarthy era when many teachers and professors were fired for their transgressions. One example was David Project's campaign against Columbia's Middle East Studies Program, that ended up on the cover of the *New York Times* and led to charges of anti-Semitism and countercharges of stoking the fire with false attacks against professors. Another is the Middle East Forum's so-called Campus Watch and the more well-known David Horowitz campaign to impose external control over higher education decision-making under the guise of "Academic Freedom" (Wiener, 2006).

Horowitz's Students for Academic Freedom movement perfectly captures the nature of the moment, critiquing schools for having too "liberal" a bias among their faculty. His group ran a national campaign to get state legislatures to pass laws that would impose external control over curriculum, hiring and firing, and teaching methods, arguing that conservatives and Christian students are being discriminated again. But like the dirty 30 list and most of these efforts, evidence of actual abuse of authority by these professors seems nonexistent. Where the efforts do succeed is in creating an environment where professors are attacked, part-time faculty are under threat of firing, and all college staff come to recognize that self-regulation is a necessity to survival and advancement.

I believe this movement is part of the aforementioned shift toward a post-panopticon form of discipline and control—a synopticon environment on college campuses, where not only the state and conservative pundits are watching professors and staff, but students themselves are being employed in the relentless attack of one of the last bastions where progressive thought is openly allowed. Essentially, the new logic has students, the media, and conservative leaders watching professors, administrations increasingly monitoring staff, and students watching each other.

The synopticon thus creates an environment not unlike the McCarthy era, where saying the wrong thing can lead to social ostracism, censure, loss of or failure to gain tenure, or firing. Given this reality, one assumes that many professors without tenure would be careful to not be too political in their classrooms, while part-time staff (who do an ever increasing proportion of all teaching) must toe the line of neutrality and apoliticism if they hope to gain renewal on their short-term contracts. In the past few years, examples abound of professors disciplined, forced to alter their classroom practice, or fired for their political stances. Among these is the case of Ward Churchill (though purportedly let go for plagiarism), Norman Finklestein not getting tenure (for his challenging of what he labeled the "Holocaust Industry"), campuses that disallowed students and professors from hanging signs from their windows or putting on bumper stickers supporting Obama in 2008, part-time firings, new campus policies on permissible faculty activity, and a whole host of other examples of limiting free speech on campus.

The result is a university that is stripped of its radical potential. Instead of critically and creatively using knowledge to explore contemporary problems or offer alternative frameworks to solve them, students are instead given knowledge elided of its political foundations. As Foucault argued, knowledge and power are inextricably linked, but if we fail to acknowledge this reality or to challenge dominant ideologies, we in fact support the status quo (Van Heertum, 2005). This appears to be exactly what conservatives like Horowitz have in mind, disabling the power of education to enlighten and empower students and provide the tools and knowledge necessary to advocate for social change. Together with the colonization of the mainstream media, the surveillance society on campuses thus has the power to undermine the last bastion of critique to the current order of things.

The power to regulate the behavior of instructors and professors, though, does not exhaust the power of the synopticon, for students are also watching each other, whether through Facebook, Twitter, YouTube, Google searches, or various other new media and technology that heighten the ability to share and gather information on others, and potentially the desire to do so. It is this last aspect that brings the synopticon into the clearest focus, and a recent example from UCLA again shows the powerful effects of media on the contemporary reality of higher education. A third-year political science major at UCLA posted a racist video on YouTube complaining about Asian students in the library. The video soon spread across campus and outraged students protested, called on the administration to intervene and even levied death threats against her. While the video was quite racist and in poor taste, the response ultimately led her to drop out of school (Daily Mail Online, 2011). A similar event occurred at Tufts in 2009, after an Asian student decided to make fun of another Asian student running for president by putting up posters with racial stereotypes, attempting to challenge what he perceived as the overly politically correct atmosphere on campus. The response of fellow students bordered on apoplectic, with protests, debates, and fears that students would be irrevocably harmed by the incident (Moltz, 2009).

These examples seem part of the reverb culture we live in today where an event occurs, is mediated, discussed, and debated, exhausted then generally tossed aside. Yet the event and its manifestations create a lever on which our social realities turn. To be is to be recorded either overtly by oneself or others or covertly through one's virtual self. Living out loud within the deterritorialized confines of cyberspace demarcates not only our commodified self but our social relations. What is not mediated in some ways does not exist. And so the desire to participate in this space manifests itself in a constant process of coactive watching, altering the nature of identity formation and interaction. Instead of simply recognizing one is being watched and altering behavior, here the process is embraced as a part of being in the new world order.

In the synopticon, people not only acquiesce to the underlying logic, they actively seek out visibility in commodified forms divorced from authentic representation. Thus the notion of the *brand* as subjectivity takes root at the ontological and social level. As Hays (2000) argues, "it is the techniques for practicing social relations, which are framed and modulated spatially, that allow for the efficient expansion of power or, alternatively, for resistance" (428).

Foucault argued that school architecture served as an intensifier of power, defining normalcy and deviance from that normalcy. Digitized data only amplifies this disciplinary power, creating dispositifs of teaching, learning, and social interaction on campuses that fit within the strictures of neoliberal ideology. Policing now goes on beyond the scope of either sovereign or disciplinary regimes—it also exists in reciprocal policing of students by professors, students by other students, professors by students, and administration of students (often in more diffuse, technological terms). Watching is thus built into the very structure of schooling and arguably alters the relationship of teachers and students to the learning process and university itself. I now turn to ID cards and their underlying logic.

ID Cards as Virtual Borders

Much of the extant research on ID cards revolves around the push for national ID cards in many countries after 9/11 (Lyon, 2004, 2007). Those debates relate, at least tangentially, to engaging discussions of borders and the issues of inclusion and exclusion. They further relate to borders both symbolic and material and thus return us to the panopticon. One could argue that if we consider ID card swiping as a sort of border (to entry, to inclusion, and to membership), then their very existence helps to demarcate education as a separate, privileged space—comparable in some ways with the idea that it is an exclusive event separated from "real life." The idea of the Ivory Tower comes to mind, along with the role of education in social sorting and ascribing of the new world order of globalization and neoliberalism in education

College IDs create eligibility markers of the campus community, much as national ID cards provide access to citizenship rights and social services. When this process becomes automatized, it can arguably undermine notions of justice and fairness just as technocratic regimes undermine debates on broader issues within the university setting, replacing them with scientific, objective analysis. As Elia Zureik argues, "By using biology and physical appearance as a means of identification, biometrics are likely to legitimate group differentiation and racialization in society in the name of security. At the extreme end of the spectrum, ethnic classification on ID cards was directly connected with twentieth-century genocide in several countries, including Rwanda" (Zureik, 2004, p. 130).

One way to look at the challenge of biometric cards is not as an instantiation of the panopticon but what Didier Bigo (2004) calls the banopticon. The argument is that ID cards do not put all under scrutiny, but single out the exceptions as quickly as possible. ID cards then essentially assign worth and risk and, within the context of the university, access to the facilities including the library. Essentially the ID card becomes a form of privilege that distinguishes the students, staff, and administration from others on campus, creating a rarified space that is not fully accessible to the population at large.

Further, it can be argued that in some ways ID cards become technologically mediated forms of objectification (Dawson, 2006). The university is increasingly overrun by corporatization and vocationalization, lending toward a proscribed notion of education as predominantly serving the economic needs of students and the larger society (Burbules and Torres, 2000; Giroux, 2004; Levy, 2006; Rhoads and Torres, 2006). As I previously argued, this supports the instrumentalization and delimiting of knowledge, undermining the broader enlightenment goals of education and the sense that higher education is about more than getting a good job. Many students, fully aware of the logic of the panopticon and synopticon, are actively driven toward this logic and their sense of themselves as brands that build an identity most attractive to future employers, friends, and potential mates. In the process, activities that cannot be readily recorded and put out into the public sphere as elements of one's virtual self become tangential and unimportant even within the basic formation of subjectivity. Thus, the higher goals of higher education become largely secondary in the shift from the credential society to a virtual identity society—with only recordable activity worthy of engaging in.

Conclusion

New forms of surveillance, together with the virtual self borne of new media and technology, are arguably moving us beyond old arguments about the panopticon toward the creation of a surveillance society where everyone is watching everyone else and themselves on a perpetual basis. This not only creates an inversion of the public and private sphere, but solidifies the neoliberal shift from citizenship based on rights and responsibilities to citizenship based on freedom of consumption, pleasure, and work.

I thus argue here that the surveillance society is altering the nature of our individual and social formation, creating new ontological realities that are fundamentally changing how we interact with the surrounding world. When one recognizes that they are under constant surveillance, they are apt to regulate their behavior to avoid being labeled as outside the normative, or worse deviant. Beyond this, the recognition that everyone is watching each other might create two contradictory tendencies—one to work to fit within the contours of the new social formation and the other to stand out and utilize those technologies to mark oneself as different. In either case, the logic of the synopticon is activated, undermining the notion of privacy and challenging the original mission of the university as an independent space for knowledge production.

The new reality is one in which higher education becomes further commodified as a key institution reinforcing the normative and social hierarchies. Technology and new media serve to inscribe the notion of the virtual self, with education largely providing one element in our data-driven, objectified public profile. The relationship of the individual to education and knowledge is thus one of skill and status acquisition rather than growth and discovery, thus fortifying the central tenets of neoliberalism and its notion of knowledge as serving purely instrumental ends—to improve opportunity (for the chosen few) and to serve the goal of social progress (defined in purely economistic terms). The surveillance society then arguably undermines the more radical and holistic notions of education, creating borders that severely constrict the role of education and knowledge in actually improving the human condition.

Bibliography

Ajana, B. (2005). Surveillance and biopolitics. *Electronic Journal of Sociology.*

Balibar, E. (2002). *Politics and the other scene.* London: Verso.

Bauman, Z. (1999). *In search of politics.* New Haven: Yale University Press.

Bigo, D. (2004). Globalized in-security: The field of the professionals of unease management and the Ban-opticon. *Traces: A Multilingual Journal of Cultural Theory, 4.*

Bourdieu, P. (1998, March 1998). L'essence du neoliberalisme. *Le Monde Diplomatique.*

Boyne, R. (2000). Post-panopticism. *Economy and Society, 20*(2), 285–288.

Burbules, N., and Torres, C. A. (2000). *Globalization and education: Critical perspectives.* New York: Routledge.

Clarke, R. (1994). The digital persona and its application to data surveillance. *The Information Society, 10*(2), 77–92.

Daily Mail Online. (2011, March 15). UCLA student who posted racist rant on YouTube says sorry as video prompts death threats. *Mail Online.* http://www.dailymail.co.uk/news/article-1366132/Alexandra-Wallace-YouTube-racist-Asian-rant-day-Japan-tsunami.html

Dawson, S. (2006). The impact of institutional surveillance technologies on student behaviour. *Surveillance & Society, 4*(1/2), 69–84.

Foucault, M. (1995). *Discipline and punish: The birth of the prison.* New York: Vintage Books.

Giroux, H. (2004). *The terror of neoliberalism: Authoritarianism & the eclipse of democracy.* London: Paradigm.

Hardt, M., and Negri, A. (2000). *Empire.* Cambridge, MA: Harvard University Press.

Hays, K. M. (2000). *Architecture theory since 1968*. Cambridge, MA: MIT Press.

hooks, b. (1994). *Teaching to transgress: Education as the practice of freedom*. New York: Routledge.

Hoy, D. E. (1986). *Foucault: A critical reader*. Oxford: Basil Blackwell.

Kim, M.-C. (2004). Surveillance technology, privacy and social control: With reference to the case of the electronic national identification card in South Korea. *International Sociology, 19*(2), 193–213.

Knight, J., Smith, R., and Sachs, J. (1990). Deconstructing hegemony. In S. J. Ball (Ed.), *Foucault and Education* (pp. 133–152). New York: Routledge.

Levy, D. (2006). Market university? *Comparative Education Review, 50*(1), 113–124.

Lewis, T. E. (2006). The school as an exceptional space: Rethinking education from the perspective of the biopedagogical. *Educational Theory, 56*(2), 159–176.

Lyon, D. (1994). *The electronic eye: The rise of surveillance society*. New York: Polity Press.

——. (2003). *Surveillance as social sorting: Privacy, risk and digital discrimination*. London: Routledge.

——. (2004). *Identity cards: Social sorting by database. OII Internet Issue Brief, No. 3*. http://www.oii.ox.ac.uk/research/publications.cfm

——. (2007). National ID cards: Crime-control, citizenship and social sorting. *Policing, 1*(1), 111–118.

Moltz, D. (2009). A tale of two posters. *Inside Higher Education*.

Piro, J. (2008). Foucault and the architecture of surveillance: Creating regimes of power in schools, shrines and society. *Educational Studies, 44*, 30–46.

Redden, J. (2000). *Snitch culture*. Venice, CA: Feral House.

Rhoads, R., and Torres, C. A. (Eds.). (2006). *The university, state and market: The political economy of globalization in the Americas*. Stanford: Stanford University Press.

Rose, N. (1999). *Powers of freedom: Reframing political thought*. Cambridge: Cambridge University Press.

Schmelzer, M. (1993). Panopticism and postmodern pedagogy. In J. Caputo and M. Yount (Eds.), *Foucault and the Critique of Institutions* (pp. 127–136). University Park: Pennsylvania State University Press.

Shapiro, G. (2003). *Archaeologies of vision: Foucault and Nietzsche on seeing and saying*. Chicago: University of Chicago Press.

Torres, C. A., and Van Heertum, R. (2011). Educational reform in the U.S. over the past 30 years: Great expectations and the fading American dream. In C. A. Torres, R. Van Heertum, andL. Olmos (Eds.), *Educating the Global Citizen: In the Shadow of Neoliberalism, 30 Years of Educational Reform in North America* (pp. 9–30). Oak Park: Bentham.

Unwin, N. B. (2009). With reference to the appropriate literature, outline the emergence of the "surveillance society." http://www.essex.ac.uk/sociology/documents/pdf/ug_journal/vol3/2009SC242_NatashaBarradellUnwin.pdf

Van Heertum, R. (2005). How objective is objectivity? A critique of current trends in educational research. *Interactions: UCLA Journal of Education and Information Studies, 1*(2).

Wiener, J. (2006, February 13). UCLA's dirty thirty. *The Nation*. http://www.thenation.com/article/uclas-dirty-thirty

Zureik, E. (2004). Governance, security and technology: The case of biometrics. *Studies in Political Economy*, 73, 113–137.

• CHAPTER SEVEN •

Socio-Technical Developments in Campus Securitization

Building and Resisting the Policing Apparatus

Ben Brucato and Luis A. Fernandez

Modern campus policing is an apparatus, or what Foucault (1980) called a *dispositif*. Campus policing involves "a thoroughly heterogeneous ensemble consisting of discourses, institutions, architectural forms, regulatory decisions, laws, administrative measures, scientific statements, philosophical, moral and philanthropic propositions . . ." (194). Further, among these heterogeneous elements, there "is a sort of interplay of shifts of position and modifications of function" (195). The specific discourses "can figure at one time as the programme of an institution, and at another it can function as a means of justifying or masking a practice which itself remains silent, or as a secondary re-interpretation of this practice, opening out for it a new field of rationality" (194–195). Campus policing "has as its major function at a given historical moment that of responding to an urgent need. The apparatus thus has a dominant strategic function" (195).

This chapter demonstrates these criteria as they apply to contemporary weapon deployment, surveillance assemblages, policing tactics, and the discursive regime they employ. In considering the range of discursive, social, technical, and infrastructural developments, we show the integrated apparatus and its constitutive elements. We also show how recent events cause the apparatus to absorb and respond to new challenges. Finally, these challenges become defined as an urgent need, the resolution of which provides a tenuous impression of security and sustains fear but at a manageable level. These provide the crucial link between the technologies of control and the broader processes of governance.

The campus policing apparatus has developed in conjunction with similar developments in municipal and federal law enforcement. Organizational and infrastructural aspects of colleges and universities have given them a particular style in constructing discursive and practical forms unique to these institutions. In this article, we detail developments that directed the particularities of contemporary campus policing strategies and connect these to broader social and discursive histories. These provide both context and motivation for particular socio-technical innovations that link surveillance, access-control, and weaponry within integrated operations and territorial control schemes. In concluding, we discuss the broader implications of these developments and identify prospects for resistance. Our point of departure is from a robust conceptualization of securitization and a discussion of the terminology related to militarism and weaponization.

The Securitization of Higher Education

Giroux has referred to the university as a "militarized knowledge factory" (2008). But what do we mean by "militarization"?

For Kraska, militarism is "a set of beliefs, values, and assumptions that stress the use of force and threat of violence as the most appropriate and efficacious means to solve problems," and it "emphasizes the exercise of military power, hardware, organization, operations, and technology as its primary problem-solving tools" (2007, 53). We agree with this characterization of militarism. However, we argue that it is more than a *mere* ideology.

What is the process of militarization, then? Again, Kraska provides some direction:

> Militarization is the implementation of the ideology, militarism. It is the process of arming, organizing, planning, training for, threatening, and sometimes implementing violent conflict. To militarize means adopting and applying the central elements of the military model to an organization or particular situation. (2007, 53)

This is a useful starting point. However, using the term "militarization" in reference to American institutions of higher learning needs to be done with considerable caution. We ought not presume a non-militarized state in an unspecified past that is somehow "in process" of becoming militarized, since the link between the military and higher education has a long history. It is best to think not of the "militarization of the university" as a transformation, but rather to uncover the practical and moral implications of an ongoing militarization.

A formal military-university connection was codified in the Morrill Land-Grant College Act of 1862, which not only granted land to state universities, but also required that campuses teach military tactics. The long history of the Reserve Officer Training Corps (ROTC) represents another college–military connection that has endured for over a century. Only during a brief time at the beginning of the 20th century did the university become idealized "as a sanctuary for independent, critical scholarship and disinterested pursuit of learning for its own sake" (Abrams, 1989, 15). During World War II, this ideal dissolved again "to accept inducements and constraints that pulled it notably away from its briefly assumed mission as a protected refuge for the dispassionate and critical study of science and society" (Abrams, 1989, 15).

We find the genesis[1] of the dispositif of militarism in contemporary campus policing in events that transpired between two primary locations. First is in the desegregation of schools and colleges that began with the University of Mississippi in 1962. Federal troops were called in to control the campus that had erupted into violence where two individuals were killed. The second is the militarized responses, generally by the National Guard, to the growth of the student movement for Civil Rights, women's rights, and against the Vietnam War. Troops occupied campuses all across the United States. At Kent State, for several days, the National Guard assaulted students with bayonets and other weapons, ultimately opening fire on May 4, 1970, killing four and wounding nine.

Given that the university has strong historical links to the military, one might instead wish to consider when the university started becoming "weaponized," which involves the introduction of policing and military weapons. Again with such a transitive term, we run into problems in demarcating the point at which a campus is weaponized from when it was not. Many technologies of control were first formed, tested, and applied at research universities. For decades, weapons have been integral to many universities. For instance, the Department of Defense has not only funded but profoundly shaped both the means and ends of scientific research at American research universities since WW II (see, e.g., Forman 1987). As a result of this long-standing military-university relationship, we cautiously use the term "weaponization" as well.

Many of the technologies under consideration in this chapter—from RFID to pepper sprays—were first invented, developed, or tested on college campuses, whether under contract by private manufacturers or directly by the Pentagon. Consider, for example, Penn State's Applied Research Laboratory, home of the Institute for Non-Lethal Defense Technologies, sponsored by U.S. Joint Non-Lethal Weapons Directorate (JNLWD). One of their projects was related to "calmatives"—weaponized tranquilizing and anti-anxiety drugs, like benzodiazepines, intended for use in pacifying unruly crowds (Lakoski, Murray, and Kenny, 2000). Penn State researchers also conduct Human Effects Advisory Panels (HEAP) to determine the effects of less lethal weapons, such as the Active Denial System (ADS), which the *Boston Globe* referred to as a "heat ray" (Kerber, 2004). Many similar technologies were developed on college campuses, first used in foreign combat, and come back to school to securitize campuses. In these cases, the chickens are literally coming home to roost. So not only do we caution against the presumption that militarization and weaponization are new, but also urge against seeing these phenomena as encroaching upon the college campus. Instead, we argue that militarization and weaponization of society is something higher education helped create.

Recent debates over (de)criminalizing students, faculty, and staff members who possess firearms on campuses in states where it is legal to possess them outside campus highlight another problem with the term "weaponization." Discussion of "weapons on campus" may carry an unspoken presumption of the legitimacy of an armed police presence at institutions of higher education. The popular and academic press regularly discuss "campus gun bans" without inclusion of law enforcement weapons. In a comprehensive review of the literature on campus weapons, Wilson and Wilson (2011) found the bulk of it "has spoken to strong opposition to allowing students to have weapons, and almost all literature dealing with the issue of arming of campus law enforcement officers has dealt not with opposition to that level of professionalization but rather with those issues related to its implementation" (33). This would suggest that when dealing with arms and weapons, it is already the accepted norm that police on campus are exempt from regulation.

Political and legal campaigns to lift police weapon exemptions have largely failed (Lipka, 2008). On the student, faculty, and staff side, there is very little empirical data on the perceived legitimacy of campus policing (Wada, Patten, and Candela, 2010). Thus, disputes over the arming of campus police are generally mute. While we see police weapons on campus as a possible location for political work by university students, discussions of weaponization in this chapter are qualified as the more thorough deployment of lethal and less-lethal weapons by police.

For reasons discussed below, we refer to the contemporary developments in the policing apparatus as *securitization*. Securitization refers to a specific technique of governance aimed at (re)producing legitimacy. In a nutshell, it works through the manufacturing of "a sudden rupture in the routinized, everyday life by fabricating an existential threat which provokes experience of the real possibility of violence and death" (Aradau, 2001).That is, ruling institutions (generally governments but in our case universities) require compliance on the part of their subjects. From time to time, a ruling body will describe a potential threat from which only this institution itself can provide security. Securitization, then, is all about securing and protecting citizens from a "threat," which provides legitimacy to undertake extreme measures to protect itself and keep its subjects secure. The terrorist attacks on the World Trade Center provided such a threat in 2001, which allowed police and global elites to connect social movements to terrorism. Similar things have occurred within universities, where outrageous but rare school shootings are invoked to substantiate the need for security. Once the threat is established, the answer comes as an increase in military-style training, weaponry, and thinking.

A "technological fix" rhetoric figures prominently in securitization discourse. The technological fix presumes that most problems can be resolved merely through employing the appropriate devices and systems. As we detail more below, highly publicized tragedies usually generate multi-million-dollar expenditures where they occurred to update gadgets and infrastructure. And these upgrades often spill beyond the location of the incident. For instance, Northern Virginia Community College (NVCC) completely revamped their surveillance, communications, and infrastructural technologies after the 2007 massacre at Virginia Tech. Yet, these technological fixes do not necessarily lead to effective systems of security. For instance, when a shooting took place at NVCC in 2009, they found 36 of 45 security cameras not working, and software used to operate them was not dependable. The emergency alert system was not used to notify students of an active shooter, and several other aspects of the NOVA Emergency Planning failed (Commission on Safety and Security, 2010). The rhetoric of the technological fix expresses the hidden presumption that past technological failings were a problem involving inappropriately or insufficient deployment of a given technology. The fix is usually towards newer, more complex devices, requiring more investment and usually resulting in more intrusion.

Securitization is partly a response to changes on a global scale. The integrated global economy is one that requires the increasingly streamlined flow of materials, bodies, and information. The function of the state is here changed. As Giroux notes, the state now "relies more heavily on its militarizing functions and the criminal justice system as a model for how to manage and contain populations within a wide range of public spheres" (2003, 59). Maintaining control of gateways and the speed with which they are traversed is now their key function. Globalization has seen the liberalization of borders, allowing resources, goods, labor, and information to more freely transcend these lines. But this has seen the concomitant rise in sophistication and intensification of protocols for surveillance and policing. As an example, rather than cutting off American universities to citizens of "unfriendly" states, these students are subjected to routine and intensive surveillance. Middle Eastern and U.S.-born Muslim students were tracked by the New York Police Department and the Central Intelligence Agency for years despite them never being suspected of any crimes. This intelligence gathering led to undercover agents joining students on rafting trips and other events, not only off their campuses but outside of the city and jurisdictional limits (Hawley, 2012).

Securitization generates considerable tension with mobility. On 9/11, U.S. Secretary of Transportation Norman Mineta responded to the World Trade Center attack as an assault on "one of the most cherished American freedoms": "the freedom of movement" (2001). Beneath the hyperbole masking a very different history, Mineta is highlighting a fundamental problematic in global capitalism: the liberalization of borders to allow the freer movement of resources, goods, bodies, and information has exposed new avenues for unprecedented risk. Transportation and communications technologies have closed distances and accelerated the movement to levels that only even newer technologies can manage. This enables what Virilio (2006) referred to as "the decolonization of the vector." As a result, control of a known threat in a known location has lost all strategic value. "All that counts is the speed of the moving body and the undetectability of its path" (2006, 151). In an increasingly mobile and accelerating world where borders become virtual, securitization requires that everyone is perceived as a potential threat, and control be conceived through rapid acknowledgment of identity and position.

These new mobilities do not eliminate borders, but liquify them. The border does not draw a line around the territory of the nation, but instead through each individual within the territory. This is not without precedent, for we see here past forms of biopower intensified with new rationale and technologies. Policing migrants and visitors, predicting crime and terror, and tracking cargo bring along with them contemporary challenges and obstacles, technological and otherwise, which provide substance for narratives of threats and the impetus and justification for the extension of ubiquitous techniques of control into daily life. The essential relationship of mobility to the most basic and routine of economic functions naturalizes securitization. In this respect, securitization is partly a response to the problematics in global capitalism. Market demands require fluidity of material and labor flows, but these present integral threats. Control over the technologies of transit and transport, identifying and monitoring bodies in motion, limiting the speed of movement, and developing choke points are contemporary socio-technic innovations central to the policing apparatus. As we see below, a variety of developments in securitization are specifically targeting mobility such that the policing apparatus can better control access and speed of movement.

Securitization generates an ambivalent discourse as its advocates employ paradoxical and opportunistic rhetoric. For example, when Pasadena City College Police Chief Peter Michael argued in favor of arming his police force, he explained that they are in the greatest position of risk, noting: "It's shown that shooters go after authority figures first" (Sum, 2010). In another example, Lt. Col. Dave Grossman, president of Killogogy Research Group, argues that the "most likely targets for attacks are believed to be middle schools where the students are not big enough to fight back and the girls can be raped" (Campus Safety, 2007). Here we see sensationalized references to situations that produce contradiction. Securitization is not about practical solutions to real problems, but rather about producing a sense of fear and the rationale for intervention by the state. What matters most is encouraging an impression of universal danger.

Constructing a Context of Fear

This section describes the discursive practices that seem to surround the securitization of the university. More specifically, we describe four types of security discourses that build the legitimacy of militarism around our colleges. These include four areas: (1) borders of legitimacy, (2) counter-terrorism strategies, (3) active-shooter response, and (4) crowd control. First, criminal policing constructs an *inside-outside* façade by using language that is focused toward crime as potentially invading the campus. Second, under the guise of "homeland security," campus policing has appropriated broader discourse of counter-terrorism that qualitatively changes the nature of the war zone they police, yet takes on localized flavor on college and university campuses. Third, the "active-shooter" response is oriented toward preventing and responding to shooting rampages like the infamous Virginia Tech massacre in 2007, emphasizing the importance of speed of response and total control of a territory. Finally, many college campuses present regular instances where police engage in crowd control, often around sporting events, turning the campus into a frequent location of intensive control of space and bodies.

Borders of Legitimacy

Criminalization is not solely a legal process, as it depends on generating a discursive divide—there are criminals, and law-abiders. With securitization, the process of criminalization is transformed to use not only *crime* but *criminals* to produce a state of danger in which publics are relieved of fear by the expanded authority and intervention by agents of the state. These agents—the police—confront a terrain as a war zone. The physical campus border becomes a dividing line of legitimacy. Campus police tend to consider "outsiders" to be a primary threat to campus security (Sloan, 1992). The discourse of campus securitization creates an inside-outside dichotomy. The broader ecology in which the campus is situated has drawn around it not only a jurisdictional border for legal and policing purposes, but also a symbolic border that defines place both geographically and within social hierarchies. The students are insiders, the legitimate bodily and cultural presence on the campus. Those who live in the surrounding community are outsiders, that potentially criminal element that must be guarded against, prevented from invading the campus with violence, drugs, and other forms of victimization of the insiders.

The inside-outside distinction brings the potential for employing symbolic codes for race- and class-based criminalization. College populations, often more affluent and whiter than the community in which the campus is situated, are to be protected from a surrounding criminality, a threat posed by the poorer, browner outsiders. This symbolic work serves to entrench existing institutionally and culturally stratifying mechanisms, and to socialize students to remain in the role of insiders. Through these discursive and policing practices, the college is not only engaging in education in fields of specialization. Additionally, they codify skin and class privileges within this role of the student. To be a good student is to shore up the inside-outside distinction, to identify as an insider, and to repel the outsider.

Students, faculty, and staff are to guard against the outsider, view any presence as suspicious—and, importantly, report such suspicious presence to the authorities. In 2011, a staff member sent a note of warning around the Rensselaer campus, stating:

> I had a black fellow, 6'tall, mid 50's black sweatshirt with red on it in our office that's suspicious. Said he's looking around for employment. PLEASE TELL PEOPLE TO LOCK EVERYTHYING. I'll pass this to my lists but please pass this to all your lists asap. Public Safety has been called. Thanks! (Brucato, 2011)

At Northern Arizona University, the Campus Police had to retract two police alerts identifying Hispanic males as assailants after the "victims" retracted their story.

The institution not only aims for students, faculty, and staff to participate in maintaining the physical and virtual border of the campus—keeping the outsiders at bay with the aid of armed police—but to behave as insiders, to be both compliant with the legal and other social processes of criminalization, but to also advance its rhetoric.

When analyzed closely, these borders relate directly to the concept of mobility, showing that the borders are symbolic and ambivalent. Ultimately, the border runs through every student as well, and when they resist, revealing their proletarian tendencies, they are likely to be met with force. The student occupation movement that emerged in light of austerity measures during the Great Recession of 2007 and the Occupy Wall Street uprising are two broad instances that saw students on multiple campuses around the country responding similarly to common crises. This resistance causes these students to be redefined as a threat, and were subsequently treated as such. Students were subjected to repression of dissent through inhibited assembly and speech, police brutality, and arbitrary arrest.

In the infamous UC–Davis incident from November 18, 2011, a significant public outcry was heard in response to videos of police brutality. These videos show several students repeatedly peppersprayed at close range while they sat on the ground during an otherwise peaceful protest. This situation demonstrates the tenuous identity of insiders, where resistance makes them subject to redefinition, and a reorientation in relation to the policing apparatus. Many of the objections to this incident were not over the presence of a militarized riot squadron, nor for the use of force *per se*, but about the "improper" use of the weapons and their use against students. When the response is to criticize the use of these policing tactics and weapons against students, the critics demonstrate the symbolic power of the inside-outside divide.

Counter-Terrorism Strategies

Much could be said here about the role of counter-terrorism in building the campus policing apparatus. We confine our analysis to three key areas. First, and most importantly, is the shifting of the discourse and strategies to the new form of *impure war* of the post-9/11 era. Next is to call attention to counter-terrorism as a key budget booster. Finally is to demonstrate the continuing and strong connections between higher education and the military-industrial complex.

For decades, police have confronted their terrain as a war zone. During the 1970s and 1980s, the drug war and the related proliferation of firearms fueled much of this tendency. Since 9/11, a "state of exception" has persisted, in which the enemy is ubiquitous but unidentified. The threat is diffused, operating as a quasi-invisible network of cells without a clear indicator of their locations. In this time of impure war (Virilio and Lotringer, 2008), sleeper cells could be anywhere, so the policing response is as though they are everywhere. These efforts are coordinated from public housing police in major metropolitan areas to rural college campus police, with the aid of the Department of Homeland Security (DHS), both instrumentally and financially.

Counter-terrorism is a key booster for law enforcement budgets (Starr, Fernandez, and Scholl, 2011), and the same is true for campus policing. In 2011, the DHS offered $2.1 billion in grants to law enforcement and other agencies to "assist states, urban areas, tribal and territorial governments, non-profit agencies, and the private sector in strengthening our nation's ability to prevent, protect, respond to and recover from terrorist attacks, major disasters and other emergencies" (Homeland Security, 2011). State and federal mandates for policing capabilities often come with grant monies to facilitate development of training, facilities, and armaments. As an example, University of Tennessee upgraded their closed-circuit surveillance system at a total cost of half a million dollars, with 40 percent coming from DHS (McCoy, 2010).

Technical developments in counter-terrorism demonstrate the bidirectional relationship between the techniques and technologies of militarization and the university. Bioweapons, chemical weapons, and aeronautical research are among key sources of revenue to research universities from the Department of Defense. Often, performing this research generates added surveillance by federal law enforcement and intelligence agencies, and, more importantly, increased access to perform it. For instance, Northern Arizona University's anthrax research has opened multiple avenues for intensive surveillance beyond the immediate setting of the laboratory. Universities are also developing counter-terrorism strategies in concert with federal agencies. The DHS "University Programs" support "critical homeland security-related research and education at U.S. colleges and universities to address high-priority, DHS-related issues and to enhance homeland security capabilities over the long term." In 2011, the DHS budgeted over $49 million for these programs (Homeland Security, 2012). Among these are the burgeoning "homeland security education" programs. Some universities offer advanced degrees in counter-terrorism, emergency management, and similar specialized programs, usually affiliated with criminal justice departments and programs.

Active-Shooter Response

After the 2007 massacre by a single gunman at Virginia Tech claimed the lives of more than 30 people, most college police departments across the country concurred that a similar situation could be repeated on their campuses. This led to a wave of policy and infrastructure changes. Most notable was the nearly universal adoption of emergency notification systems. These communications networks were designed to rapidly notify all students, faculty, and staff in the event of a threat to security and safety on the campus via email, text messages, public announcement, and other technologies. This might be compared with Cold War–era air raid drills, typified by the infamous "Duck & Cover" propaganda videos. The Federal Civil Defense Administration (FCDA) administered programs of "preparedness" and "emotional management" to respond to the constructed threat of imminent nuclear warfare in the early Cold War period. These same themes of preparedness and emotion management run through the discourse of the active-shooter response.

In addition to the notification systems, campus lockdown tactics were developed, boosting specialized consulting and instruction programs to update officer training. These response systems integrate notification systems with policing tactics and command center operations to attempt complete control of an area in which an active threat is determined. Perhaps most notable was the intensification of surveillance with the building concern over active-shooter scenarios. Building access control was heightened, often using radio-frequency identification (RFID) cards to limit and monitor entry. Some universities additionally control access to individual rooms within buildings. Nonetheless, three years after the Virginia Tech incident, one-third of campus police departments report that they are unprepared to respond to an active-shooter scenario, and "two in five say their institutions don't dedicate enough money, resources and personnel to campus safety and security efforts and technology" (Gray, 2010). A single event was used to generate a pervasive sense of fear of an imminent threat for a repeat of this event at any of the more than 6,600 colleges and universities in the country. It is a forceful rationale for increasing the power of the policing apparatus, both for the police and the policed.

Crowd Control

Many universities and colleges regularly hold sporting events on campus that draw crowds from the hundreds to the tens of thousands. On one hand, we see preparation, readiness, and the integration of techniques and technologies by police. Campus police and security officers are technologically equipped and tactically prepared for crowd-control scenarios. These events bring together the full use of surveillance, access control, operations management, and a range of weaponry discussed below. Additionally, they shore up connections among campus police with other jurisdictions for coordinated policing of large, heterogeneous crowds.

On the other hand, we see the normalization of a visible presence of militarized policing, the restriction of movement, the submission to surveillance, and all these merging with leisure activities on campus. The regular deployment of such crowd-control technologies and tactics generate experience and readiness among police and security forces. This goes hand in hand with the normalization of the deployment of weapons, barricades, riot squadrons, and other symbols of control that would otherwise seem intrusive in an institution of higher learning and antithetical to widely held pedagogical values. The routinization and visibility of crowd-control tactics build, demonstrate, and reproduce the legitimacy of militarized policing, both symbolically and practically.

The preparedness for routine crowd-control scenarios shows capacity of campus police for handling protests and uprisings. When Occupy Los Angeles was confronted by hundreds of riot police at the University of California–Los Angeles, many of them on the UCLA police force (Romero, 2011), this represented only a subtle shift from routine crowd-control strategies. However, like in the broader context, police more rapidly engage politicized crowds, make more arrests, and do so with more brutal tactics. Sports riots that result in considerable property damage and cause widespread injuries generally lead to few arrests and even fewer charges when compared with protests and other political events. This is not without exception. Consider the case of John McKenna. He was arbitrarily singled out of the crowd, thrown into a concrete wall, and struck with a baton during celebrations following a University of

Maryland basketball game. McKenna sustained a concussion, and a head wound that had to be closed with staples. For this, he was charged with assaulting an officer. Nearby surveillance cameras captured footage of the entire incident, and McKenna's charges were subsequently dropped and two officers were charged with assault (Castaneda and Zapotosky, 2010).

Generally, these social and discursive developments are reactive. In the case of the counter-terrorism strategies and "homeland security" discourse, the state and other institutions of power—including those of higher education—are responding to the innovations of a mobile, quasi-invisible, and diffused resistance to global hegemony that take myriad forms from Salafi jihadists to the Earth Liberation and Animal Liberation Fronts. The construction of "crime waves," whether by drug rings or serial rapists, merge with the threat of "active shooters" or "active bombers" to represent an unidentifiable and abstract criminal element to be prevented from invading the tranquil safe space of the campus and victimizing insiders. These go beyond any sort of moral panic, using real cases to instill a pervasive and intense sense of fear to be assuaged by the deep intervention into campus life by police. This is a shock doctrine (Klein, 2007) de-territorialized—where a crisis on one campus is a threat to all campuses—and resolved with the technological fix of the policing apparatus. These shocks and their subsequent reactions are met with waves of new regulations and funding from federal agencies that advance the incursion of the policing apparatus into the campus space. Sometimes, however, the expansion of policing technologies is charged directly to the students. Providing security guards with handguns at Evangel University in Missouri resulted in students being charged an additional annual $10 "security fee" (Campus Safety, 2010).

Socio-Technical Developments: Integrated Operations and Territorial Control Schemes

The college campus provides unique advantages to policing strategies. Municipal policing encounters multiple conflicts over the private-public divide. The campus presents infrastructural and policy capabilities that allow a distinctive synergism. A control ecology is built through the integration of surveillance, communications, and weaponry into a comprehensive infrastructure. Campus policing experts advocate the use of operations centers, which one describes as

> a critical component to the development of the master security plan. The [Security Operations Center] is a 24x7 command, control and security operations management area. This is the primary hub for operational communications, information sharing and situational awareness for all information pertaining to campus activities, threats and emergencies. (Fields, 2009)

The operations centers manage "an ever-changing onslaught of daily threats and warnings from security and enterprise systems." Importantly, they are to respond definitely to the question "How does one categorize and prioritize threats?" (Fields, 2009). This theme of categorization is one we will return to below.

These systems of information management are intimately connected to weapon deployment and policing tactics, especially in the cases of campus lockdowns and crowd-control situations. This degree of technical, organizational, and infrastructural synergy is uncommon. Similar examples exist in large public infrastructures, like mass transit stations, and "total institutions," like prisons and hospitals. Here we look at three clusters of technologies that are integrated in these control ecologies: surveillance and communications, access control, and weaponry.

Surveillance and Communications Technologies

Campus policing involves managing, responding to, and communicating information, and this integrates a number of devices for communications and surveillance. Monahan defines surveillance systems broadly to mean "those that afford control of people through the identification, tracking, monitoring, or analysis of individuals, data, or systems" (2010, 8). For this reason, surveillance is a systematized expression of power. Foucault explained that power is embedded in social processes and advances through "a set of practices that organizes individuals and the spaces they occupy for the production of docile bodies" (Simmons, 2010, 55). These practices are myriad, embedded throughout institutions and particularly in schools. Among the more apparent forms of control in the production of docility are the security cameras, identification documents and routines, and a range of electronic surveillance machines and the computers that manage the data collected that populate the modern landscape. Simmons explains how technologies have allowed for softer forms of discipline and control in producing docile bodies in the "school space." Surveillance is designed for "managing students and orchestrating their daily engagements" (2010, 55). Curriculum alone represents an expression of modern power, she argues, but these "everyday expressions of control and governance [do] not diminish the significance of high-tech control networks and surveillance systems . . ." (Simmons, 2010, 55).

For Foucault, this relationship is not solely repressive. Security is aimed at producing productive bodies, directing their activity, and shaping what they produce, which allows for some organic work, but within specific parameters. Engaging with the surveillance apparatus as a student is a form of training not merely to comply and submit one's body to be viewed, tracked, and controlled. It defines the parameters for acceptable activities, demarcating the lines within which normative, productive acts can be performed. Moreover, this process is instructive of a logic of engagement within a specific system of production that will be redeployed by students later in their professional lives, often with the roles reversed in relation to the power of the surveillant apparatus. Many will eventually manage employees through surveillance techniques and devices, monitoring their activities and tracking their movements.

Perhaps the most common impression people have of surveillance technology is of video cameras and CCTV systems. This impression holds true on the college campus, as cameras are often ubiquitous. Breakthroughs in high-definition (HD) technologies have led to recent deployment of new camera systems. For instance, at the University of Miami, over 500 networked HD megapixel cameras using H.264 Main Profile compression are in use (Campus Safety, 2012). Rising concern over lone-shooter response boosted the use of surveillance cameras inside buildings and on campus grounds. Beyond videographic surveillance, institutions of higher education survey telephone lines and email accounts, using software and direct monitoring. Some use RFID and other technologies to monitor presence in buildings or classrooms. Most important to these devices are their integration in networks to manage data in singular databases, compiling records for locations and profiles for students.

The campus is not only a location for the deployment of surveillance technologies. Many breakthroughs in new surveillance technologies originate in campus laboratories. University research in robotics, such as the Wyss Institute for Biologically Inspired Engineering at Harvard University, is taking new surveillance technologies in new directions. The Wyss Institute's Kilobot system "includes robots about the size of a quarter, a controller to program and manage all the robots at once, and software. The design allows a single user to easily oversee the operation of a large Kilobot collective including programming, powering on, and charging all robots, all of which would be difficult, if not impossible, using existing robotic systems" (Wyss Institute, 2011). While primarily aimed at search and rescue applications, additional surveillance functions are likely to creep into use as the devices gain greater diffusion.

Surveillance involves the classification and sorting of subjects. Security expert Jeff Fields provides the rationale behind this:

> Since the SOC [Security Operations Center] is the heart of the physical security operation, campuses should be equipped with state-of-the-art communication and operational equipment. In addition to control, operators should have the ability to classify using integrated access, intrusion and video technologies. Historically, SOCs have been designed to extract information by visually monitoring the multitude of cameras, readers and intrusion devices. The objective for today's campus physical security master plan is to have data pushed to the SOC based on condition, severity and protocol. The increase in response, consistency and accuracy can make the difference in a variety of situations throughout the campus. (2009)

However, David Lyon (2003) describes this process as "digital discrimination," in which "surveillance not only rationalizes but also automates the process" (13) of categorization, increasingly performed by computer code, which expands the scope and accelerates the process of sorting. These categories become obdurate, and "values, opinions, and rhetoric are frozen into codes" (Bowker and Starr, 1999, 135). This quite literally becomes an automated and autonomous mechanism to reinforce class, race, and other divisions. This point alone is reason enough for a careful attention to the technologies in play in the militarization of higher education. Much like legislative acts, technologies govern behavior and bodies. "Technological processes in contemporary society have become the equivalent of a form of law," writes James Carroll, "that is, an authoritative or binding expression of social norms and values from which the individual or a group may have no immediate recourse" (in Sclove, 1995,11). That this is the case in campus surveillance is hardly debatable—security experts advocate this as a "best practice."

Surveillance technologies are designed as a one-way form of communication, conveying information such as identity, position, time, and other details to a remote, unknown viewer. Mass notification systems reverse this direction. The source here is invisible, often a system of codes triggering mass alerts using loudspeakers, text messages, and voice calls. These systems aim to instantaneously send a concise message of threat to all students, faculty, and staff members on and off campus, and anyone else present in the vicinity. These elaborate systems combine computer and telephony software and hardware, network switches and routers, public announcement systems, and other devices. These systems come at a considerable commitment of funds for material capital and labor for installation and maintenance. In doing so, these solidify into the material relations the perception of imminent danger and threat.

Mass notification systems are federally mandated. The U.S. Department of Education (DOE) determined that Virginia Tech was too slow to notify students during the shooting in 2007. The university was fined $55,000, and the DOE considered pulling $98 million in federal student aid (D. Potter, 2011). Here we see government regulation and its application creating waves of reactive developments on campuses across the country. Subsequently, campuses nationwide spent millions updating campus notification systems, including automated text message and telephone alert, public announcement, and other related systems.

These surveillance and communications technologies are integrated into the overall infrastructure and operations of the entire campus. In this respect, the ability for policing to colonize the architecture and geography lack many parallels. Despite the largely decentralized relationships within institutions of higher learning, the discourse and practice of securitization has streamlined the ability of the police to, for instance, put a camera wherever they want. Perhaps more concerning is that campus police have been able to be opaque about who is collecting what information and how it will be used.

Access Control and Speed Limits

The control of access to buildings and rooms using electronic locks connected to RFID scanners allows not only the management of access, but the immediate acknowledgment of identity and position by databases and the operations officers who monitor activity for policing. Turnstiles are advocated to slow movement in deliberate choke points. Increasingly, entryways are being controlled using biometric readers, like the Hand Key "hand geometry reader" that "simultaneously analyzes more than 31,000 points and instantaneously records more than 90 separate measurements of an individual's hand, including length, width, thickness and surface area, to verify the person's identification" (Brigham, 2009).

Cleveland State University director of security credits the use of turnstiles with having "completely eliminated the possibility of getting through [entryways] unaccounted for" (in Chamberlain, 2011). Some universities are integrating access point controls with databases to monitor students over time. A director of operations at Arizona State University explains their monitoring program:

> . . . if a student hasn't been to class recently, we can check and see when and if he's been to his apartment . . . . If he has been up there but not attending class, counselors can then contact the student to determine whether or not there's something serious, emotionally or otherwise, going on with him. If he hasn't been up to the room at all, we can alert the proper people. (in Chamberlin, 2011)

Virilio (2006) adds an important dimension to Foucault's (1980) notion of biopower, where not only are bodies to be controlled using biometric and other identifiers. Additionally, the speed with which these bodies move through the terrain must be monitored and controlled. The neoliberal world requires mobility, not just for the purposes of material production, but for the production of minds and bodies that can integrate into its controlled ecology. But this ecology is controlled through the management and recording of access and the limiting of speed. In this regard, technologies for controlling access and limiting speed of movement are intrinsically wedded to the logics of surveillance—the goal is not merely to witness and record movement, but to control and limit its direction and pacing. Foucault was concerned with the control of the body, Virilio with the control of speed. Together, they help connect the merger of surveillance and access control on campuses with broader processes of governance.

Surveillance, communications, and access control coalesce around important functions in campus lockdowns and crowd control. When the ubiquitous, evasive, invisible, and mobile threat is located and identified, the confluence of surveillance, notification, access control and operations technologies are integrated with new lockdown tactics to isolate and neutralize it. The campus lockdown demonstrates battlefield militarism plainly without even considering the involvement of weaponry.

Weaponry

In the 1970s, the growth of automobile usage and of student protest movements drove a redefinition of policing roles and approaches. Student protest movements led police to focus on crowd control, driving tactical and technological change in campus policing. Up until the mid-1940s, any campus security guards had a "watchman-janitor image" (Gelber, 1972). By the 1970s, however, many campuses developed their own police departments. In Gelber's research for the U.S. Department of Justice (DOJ), he found very few of these departments were developed for a perceived risk of crime or any noted incidents. Half of the police department directors he surveyed indicated the increased use of automobiles on campus drove the shift from the "watchman-guard" function to one of "traffic and crime control" (1972, 29). In the early 1970s, about half of the states allowed state university police departments to make arrests, and only seven states allowed police at private institutions to do so (1972, 35). At this time, only 26 states required mandatory training for campus security. So, in the early 1970s, the modern police department had not yet taken hold on campuses.

This DOJ report demonstrates a shift in police functions and the tools they employ. Gelber wrote:

> The availability of chemical properties such as tear gas, mace, and pepper fogger to subdue unruly campus crowds raises many philosophical questions concerning the relationship between security officer and student but from a policing point of view it is also significant. (Gelber, 1972, 80)

According to Wright (1978), during the 1970s, governments had invested "huge sums on the research, development, and procurement of new equipment for their police forces. The resultant innovations in police technology have been functionally designed to facilitate an extension of the scope, efficiency, and growth of police power" (305). Early in the decade, these had yet to take hold on college and university campuses. He found that 64 percent of private colleges and 36 percent of public colleges reported no crowd-control equipment available (Gelber, 1972, 80). Nonetheless, a strong majority of campuses had training and strategies in place for "mass disorder," most often involving municipal police aid (Gelber, 1972, 124–132). The student protest movements of the late 1960s and early 1970s were clearly generating considerable attention by college and university administrators, representatives in government, and police agencies. This was reflected in prioritizing crowd-control strategies and a focus toward further deployment of less lethal weapons.

Four decades later, we see the rapid growth of the less lethal weapons industry, in what Arike aptly described as "what appears to be the first arms race in which the opponent is the general population" (2010). Through 2009, the Joint Non-Lethal Weapons Program (JNLWP) alone received $817 million in public funds to develop new nonlethal weapons for military and police use (Government Accountability Office, 2009). "The trend is now away from chemical and 'kinetic' weapons that rely on physical trauma and toward post-kinetic weapons that, as researchers put it, 'induce behavioral modification' more discreetly" (Arike, 2010). The Homeland Security Research Corporation determined that the less lethal weapon market will triple by the end of the decade (2011).

This market is opening up new innovations in weaponry. Taser International recently announced the double-barreled Taser X2 ECD that "features a second shot for instant-miss recovery, dual lasers for improved accuracy, an enhanced power magazine enabling more than 500 firings and a warning arc to help keep conflicts from escalating. Additional features of the X2 include semi-automatic back-up shot capability for multiple targets, improved weather resistance and self-diagnosis features" (Campus Safety, 2011).

The name less lethal implies the minimization of potential deaths. However, Everette Howard, an African American summer-school student at the University of Cincinnati, died after being Tasered by a campus police officer (Caulfield, 2011). Howard is one of hundreds of unarmed people killed with Tasers used by police (Amnesty International, 2008). But aside from their alleged safety, another purpose behind less lethal weapon use is to allow for more degrees of force to be used and to enhance opportunities for police officers to escalate force. Because less lethal weapons have a lower threat threshold for use in the subduing or apprehension of suspects, police are more frequently able to physically harm them within the scope of "use-of-force" guidelines.

When Gelber was conducting research for the DOJ in the early 1970s, campus police armed with guns were all but absent. Currently, though, about two-thirds of campuses have an armed police force (Reaves, 2008), and recent data shows this trending upward (McBride, 2009). Ninety-one percent of campuses with a student population over 15,000 are weaponized (Reaves, 2008).

One tack on advocacy of arming a campus police force is exemplified by the rhetoric of self-protection: the unarmed police officer is a liability risk to the institution by not having the proper equipment to be safe. Writing for industry publication *Campus Safety*, Shad Ahmed compared a doctor's stethoscope to an EMT's glove and mask. The former is an example of "tools of the trade," while the latter is "personal protective equipment" (PPE). "Many people do not realize that the use of a firearm can be (as is most often the case) considered PPE and not just a tool of the trade," he writes (Ahmed, 2009). He explains standard firearm certification programs

> train their officers to use firearms as a primary form of PPE. Officers are taught to unholster and/or draw their weapon for certain situations when their personal safety could be compromised. It is important to note that this does not always mean that a weapon is pointed at someone and even less so that shots are actually fired. It is a deterrent and PPE just in case it is needed. Just like the EMT's gloves: it may not always be necessary but when it is, the EMT is glad to have worn them for their own safety.

The very weapons police claim make them safer have been used to unduly threaten the safety of others. Consider the case of Kofi Adu-Brempong and others like him. Police responded to a noise complaint and burst through the door to the campus residence of Adu-Brempong, a Ghana native and graduate student at the University of Florida. Campus police shot him in the face (Stanfill, 2010), perhaps because they confused him for an "outsider."

Resisting the Policing Apparatus

Resisting the campus policing apparatus is essential and an ongoing effort. We conclude the chapter with some thoughts on resistance, painting broad strokes of what is possible. A key benefit of tracing the genesis and development of a dispositif, or apparatus, is to locate key opportunities for effective resistance. The military–university relationship began with the land-grant system, and historic challenges to the campus presence of the Reserve Officer Training Corps (ROTC) serve as an example of connecting contemporary resistance to the point of origin for the problems we face today. The modernization of campus policing originated with military invasion of campuses where students rose up against war and for social justice. This provided a key rationale for the expansion of the roles for police, and for increased funding, weaponization, and authority. Student uprisings are a primary cause of campus policing innovations, and therefore proactive innovation that exposes and threatens existing weakness—those popular innovations the apparatus has not yet adjusted for—pose an opportunity for resistance. Past examples of this can be seen in animal rights and environmental activists who clandestinely destroyed laboratory and other equipment, barred access to buildings, and undermined research projects that exploited nonhuman animals. We are not advocating the recreation of tactics, but to look to these past examples to what made them successful: the campus policing apparatus did not anticipate and were not prepared for innovative modes of resistance. This class of innovation by student activists is of the type that the genealogy of the apparatus points toward for effective resistance.

As we have already mentioned, securitization is a still-active process, not a completed event of the past. On the one hand, this continual securitizing process signals that control systems continue to innovate in response to sites of resistance (cf. Hardt and Negri, 2005). On the other hand, they are also reactions to catastrophic breakdowns in the social order. An increasingly common example of the latter point are of those alienated young men who take to massacres or suicide-by-cop, in either event providing a living example for the ubiquitous threat of subjective violence. At UC–Berkeley, non-student activists linked up with student activists for an Occupy Berkeley event (a direct challenge to the inside-outside divide). This resulted in widespread police brutality that was captured on video and broadcast around the world. Critically important here is that this same day, a student brought a gun into a building on the same campus, as an alleged staging for a successful suicide-by-cop. In a single day on one campus, we see both kinds of events. While Hardt and Negri (2009) may overstate the commonality in the direction of influence, we clearly recognize that many of the developments we have discussed are reactive innovations. But likewise, we recognize that some of the events that spur securitization are a result of ruptures and integral accidents. The result of this confluence of influences is the creation of a thought style that sees an enemy everywhere, and this enemy is mobile and elusive, to be universally feared. Most important, the state and its police aim to generate the legitimacy to do whatever is necessary to ensure "safety" and "security" for some, while allowing for mass inequality to remain as the status quo. We must challenge and fight this.

At the same time, we see in this history an important discursive development. The apparatus viciously attacks effective innovators—animal rights activists and environmentalists who identified and attacked weak points in the system are now branded terrorists. This labeling goes beyond a rhetorical tool, as policing responses to these activists rival the Red Scare. Will Potter (2011) has chronicled this "Green Scare." Challenging the counter-terrorism discourse becomes yet another crucial site of resistance, as Salter (2011) writes:

> Resisting and actively undermining the rhetoric of terrorism, the associating and framing of dissent as *terrorist-like*, limits the ability of the State and corporate interests to label activists in such ways, enabling dissent and the renewal of revolutionary efforts aimed at moving society towards an existence free from multiple fronts of systematic oppression: a society embracing total liberation and true liberatory politics. Such actions are an essential element of living and resisting, justly, in a repressive society. (234)

We here underscore that this is *an* essential element, and that this brings *justice*—not *success*—to our resistance.

Another key locus for resistance is in the relationship between research laboratories and the corporations and military, particularly in developing new technologies of repression. Here, students can not only build solidarity with the millions of people assaulted by imperialist wars and the internal colonialism of police, but they can also preemptively confront those technologies in the development stage, those that have historically been brought back to the campus for use by police against students. U.S. research universities are developing biometrics and robotics for use in surveillance, enhancements for range and guidance in ballistics and aeronautics, and technologies to manage mobility. Further, the college and university is a point of production of those who will continue these developments for corporations and militaries. These present multiple locations for potential effective resistance.

The process of securitization shows a trend toward policing that perceives city spaces as areas of conflict, populated by shadow enemies. This relies on an inside-outside rhetoric that defines crime as an off-campus phenomena and the campus as guarded against invasion. But this rhetorical divide is made material through the policing of jurisdictional borders, profiling of the poor and people of color in and around campus. Further, it is embedded in codes for categorization in surveillance and operations protocols, appropriated into the built environment as access controls and monitoring equipment are integrated into the architecture and grounds. On the other hand, the inside-outside divide breaks down quickly, as insiders are subject to redefinition by police. This symbolic order is easily accepted, and insiders may have weapons turned on them. As Graham (2010) shows, this symbolic order is being "radically reimagined." Citing Feldman, Graham discusses the "securocratic wars" that have led to ubiquitous borders, within which "open-ended, deterritorialized wars (on drugs, crime, terror, illegal immigration, biological threats) organized around vague, all-encompassing notions of public safety rather than around territorial conquest. Their purpose is to maintain state sovereignty, not through external war combined with internal policing, but through raising the specter of mobilities and flows deemed to contaminate societies and threaten the social order, both internally and externally" (2010, 91).

This inside-outside divide identifies an important site of resistance. Building solidarity across the divide requires the legitimacy of police, their tactics, and tools be fundamentally challenged. When student activists demand a free-speech safe zone, and challenge police incursions on this space by police violence, it is important that they also connect with the adjacent community, which is likely subjected to violence and discrimination much more frequently.

Discursively, we point to the brief but meaningful moment when the universities were understood as locations for knowledge, safe spaces from the "broader" society. While this strategy has obvious class-based problems, we think that students could still make claims for weapons-free campus, more and better healthcare, and a broader set of social services. However, what would make this strategy radical is not the claim, but rather its link for the same benefits across the inside-outside divide. That is, making the claims universal, not only for students.

Unfortunately, the securitization process that finds "threats" to justify police management, budgets, and armament is likely to continue. Campus policing has lagged behind municipalities, but the proliferation of less lethal weapons and firearms in the hands of campus police, as well as their training as militarized riot squadrons are visible markers that they are catching up. Given the economic crisis resulting in cuts in higher education and social services in general, we are likely at the start of a more volatile historical period that will include mass mobilizations like the ones we saw recently with the Occupy Movement. As students and faculty, we must pay attention to how the potential "threat" for mobilization and instability will be constructed to justify increases in police powers. The Occupy Movement has already showed us a strategy, pointing out that much of what passes for security is really about keeping the 1 percent safe, while leaving the rest of us paying the bills.

The university is also an important location to study because it is a microcosm of the larger society. As such, close scrutiny of what happens in the laboratory, institutes, and police departments on campus will likely indicate innovations of the police apparatus. Thus, confronting them here might be strategically smart, helping to prevent them from spreading to the larger society. As a microcosm of the broader society, the university already shows us some initial insights. First, we can see that control is contradictory and it is applied unevenly. This can be seen across universities, where public universities experience a different kind of control than private ones. We can also see the uneven configuration in the inside-outside distinction, where communities of color that sometimes develop on the fringes of large institutions are policed with harshness, particularly around drug-related issues. Yet, on the "inside," students might be ticketed (rather than arrested or imprisoned) for similar drug behavior. Again, these contradictions need to be confronted through solidarity that crosses boundaries.

We can also clearly see the university already innovating and leading the way on surveillance for public order maintenance. We already know from various examples (e.g., CCTV, biometrics) that surveillance is one of the leading modes of control for the future. And these systems are being designed in and through the university settings, with some applications already in place. For instance, Northern Arizona University (among others) is already experimenting with student movement monitoring through their university identification cards (Brucato, forthcoming). It is crucial that students and faculty start thinking and be actively engaged in uncovering the resistance possibilities. Fortunately for all of us, there are people already thinking and acting out. We look forward to seeing what innovations they come up with next.

Note

1. In Foucault's approach to the *dispositif*, it is necessity to locate its genesis. This extends beyond the important methodological implications to also highlight opportunities for resistance. While the above says that we are unable to precisely locate the origin of *the military–university relationship*, we can locate the genesis of militarism in *contemporary campus policing* in these events. As we will demonstrate below, prior to these events, campus police fulfilled a significantly different function on campus, and almost no campus police were armed. The socio-technical developments discussed below depended upon these and similar events, and the discourse that surrounded and emerged from them. Additionally, historic sites of resistance are located here. For instance, the National Guard first confronted Kent State students during an uprising against the ROTC, two days before the massacre. The Mississippi incident was a site of crossing the *inside-outside* divide (discussed later in the chapter) on social justice issues, particularly desegregation.

Bibliography

Abrams, R. M. (1989). The U. S. military and higher education: A brief history. *Annals of the American Academy of Political and Social Science*, 502: 15–28.

Adams, R. F., Ayer, A. C., and Bishop, L. R. (2011). Arming campus police: Managing the risk. *University Risk Management and Insurance Association Journal*,9–17.

Ahmed, S. U. (2009, May 1). The lethal question: To arm or not to arm? *Campus Safety*. http://www.campussafetymagazine.com/Channel/Public-Safety/articles/2009/05/The-Lethal-Question-To-Arm-or-Not-to-Arm.aspx

Amnesty International. (2008). "*Less than lethal*"? *The use of stun weapons in U.S. law enforcement*. London: Amnesty International Publications.

Aradau, C. (2001). Beyond good and evil: Ethics and securitization/desecuritization techniques. *Rubikon*. http://venus.ci.uw.edu.pl/~rubikon/forum/claudia2.htm

Arike, A. (2010, March). The soft kill solution: New frontiers in pain compliance. *Harper's Magazine*, 38–47.

Bowker, G., and Starr, S. L. (1999). *Sorting things out: Classification and its consequences*. Cambridge: MIT Press.

Brigham, T. (2009, November 1). Rec center gives biometric readers a workout. *Campus Safety*. http://www.campussafetymagazine.com/Channel/Security-Technology/articles/2009/11/Rec-Center-Gives-Biometric-Readers-a-Workout.aspx

Brucato, B. (2011). Email received from RPI faculty and student list-serv.

——. (forthcoming). From accountability policy to surveillance practice. In K. Ball and L. Snider

(Eds.), *The Surveillance-Industrial Complex*. New York: Routledge.

Campus Safety. (2007, September 29). U.S. schools not prepared for terrorism, experts say. *Campus Safety*. http://www.campussafetymagazine.com/Channel/Security-Technology/News/2007/09/29/US-Schools-Not-Prepared-For-Terrorism-Experts-Say.aspx

Campus Safety. (2010, April 30). Missouri College arms security officers. *Campus Safety*. http://www.campussafetymagazine.com/Channel/University-Security/News/2010/04/30/Missouri-College-Arms-Security-Officers.aspx

Campus Safety. (2011, June 21). Double-barreled TASER X2 ECD. *Campus Safety*. http://www.campussafetymagazine.com/Channel/Public-Safety/Products/2011/06/Double-Barreled-TASER-X2-ECD.aspx

Campus Safety. (2012, January 21). U. of Miami expands campus video surveillance. *Campus Safety*. http://www.campussafetymagazine.com/Channel/University-Security/news/2012/01/21/U-of-Miami-Expands-Campus-Video-Surveillance.aspx

Castaneda, R., and Zapotosky, M. (2010, September 20). 2 Prince George's officers indicted in beating of University of Maryland student. *Washington Post*. http://www.washingtonpost.com/local/xxxx/2011/09/20/gIQAMYbviK_story.html

Caulfield, P. (2011, August 8). Student, Everette Howard, 18, dies after being shot with Taser by campus cops at U. Cincinnati. *New York Daily News*. http://www.nydailynews.com/news/national/student-everette-howard-18-dies-shot-taser-campus-cops-u-cincinnati-article-1.947118

Chamberlain, B. (2011, July 8) How turnstiles can protect your campus. *Campus Safety*. http://www.campussafetymagazine.com/Channel/University-Security/articles/2011/07/How-Turnstiles-Can-Protect-Your-Campus.aspx

Commission on Safety and Security. (2010). Report of the presidential commission on safety and security. Report published by Northern Virginia Community College.

Fernandez, L. (2008). *Policing dissent: Social control and the anti-globalization movement*. Brunswick, NJ: Rutgers University Press.

Fernandez, L., and Joel Olson. (2011). To live, love, and work anywhere you please: Arizona and the struggle for locomotion.*Contemporary Political Theory*,*10*: 412–419.

Fields, J. W. (2009, March 1). 10 steps to creating acampus security master plan. *Campus Safety*. http://www.campussafetymagazine.com/Channel/Security-Technology/Articles/2009/03/10-Steps-to-Creating-A-Campus-Security-Master-Plan.aspx

Forman, P. (1987). Behind quantum electronics: National security as basis for physical research in the United States, 1940–1960. *Historical Studies in the Physical and Biological Sciences*, *18*(1): 149–229.

Foucault, M. (1980). *Power/knowledge: Selected interviews & other writings, 1972–1977*. New York: Pantheon.

Gelber, S. (1972). *The role of campus security in the college setting*. Washington DC: U.S. Government Printing Office.

Giroux, H. (2003). Zero tolerance, domestic militarization, and the war against youth. *Social Justice*, 30(2): 59–65.

——. (2008). The militarization of U.S. higher education after 9/11. *Theory, Culture & Society*, *25*(5): 56–82.

Government Accountability Office. (2009). *DOD Needs to improve program management, policy, and testing to enhance ability to field operationally useful non-lethal weapons*. United States Government Accountability Office: Washington, DC.

Graham, S. (2010). *Cities under siege: The new military urbanism*. New York: Verso.

Gray, R. H. (2010, November 2). How safe is your campus? Survey results: Universities. *Campus Safety*. http://www.campussafetymagazine.com/Channel/University-Security/Articles/2010/11/University-Public-Safety-Statistics.aspx.

Hardt, M., and Negri, A. (2005). *Multitude: War and democracy in the age of empire*. New York: Penguin Books.

——. (2009). *Commonwealth*. Cambridge: Belknap Press.

Hawley, C. (2012, February 20). NYPD monitored Muslim students all over northeast, *Yahoo! News*. http://news.yahoo.com/nypd-monitored-muslim-students-over-northeast-201653480.html

Homeland Security. (2011, May 19). DHS announces grant guidance for fiscal year (FY) 2011 preparedness grants. *Homeland Security*. http://www.dhs.gov/ynews/releases/pr_1305812474325.shtm

Homeland Security. (2012). FY 2012 budget-in-brief. Report published by the U.S. Department of Homeland Security.

Homeland Security Research Corporation. (2011). Non-lethal weapons: technologies & global market, 2012–2020. Report published by HSRC.

Kerber, R. (2004, September 24). Ray gun, sci-fi staple, meets reality. *Boston Globe*. http://www.boston.com/news/science/articles/2004/09/24/ray_gun_sci_fi_staple_meets_reality?pg=full

Klein, N. (2007). *The shock doctrine: The rise of disaster capitalism*. New York: Picador.

Kraska, P. B. (2007). Militarization and policing: Its relevance to 21st-century police. *Policing, 1*(4). 501–513.

Lakoski, J. M., Murray, W. B., and Kenny, J. M. (2000). *The Advantages and limitations of calmatives for use as a non-lethal technique*. Report published by Penn State College of Medicine, Applied Research Laboratory. http://www.mindfully.org/Reform/Non-Lethal-Calmatives3oct00.htm

Lipka, S. (2008, April 18). Campaigns to overrule campus gun bans have failed in many states. *Chronicle of Higher Education*.http://chronicle.com/weekly/v54/i32/32a01601.htm

Lyon, D. (2003). Surveillance as social sorting: Computer codes and mobile bodies. In D. Lyon (Ed.), *Surveillance as social sorting: Privacy, risk, and digital discrimination*. New York: Routledge.

McBride, J. (2009). Should campus police be armed? *Law and Order, 57*(6): 62–64.

McCoy, L. X. (2010, September 9). New cameras to keep watch over Neyland Stadium.*Knoxville News Sentinel*. http://www.knoxnews.com/news/2010/sep/09/new-cameras-to-keep-watch-over-neyland-stadium/

Mineta, N.Y. (2001, September 11). Statement of Secretary of Transportation Norman Y. Mineta. *The Avalon Project: Documents in Law, History and Dipolomacy*. http://avalon.law.yale.edu/sept11/dot_005.asp

Monahan, T. (2010). *Surveillance in the time of insecurity*. New Brunswick, NJ: Rutgers University Press.

Potter, D. (2011, March 30). Virginia Tech fined $55K for response to shootings. *Washington Post*. http://www.washingtontimes.com/news/2011/mar/30/Virginia-Tech-fined-55000-shootings/?page=all

Potter, W. (2011). *Green is the new red: An inside's account of a social movement under siege*. San Francisco: City Lights Books.

Reaves, B. A. (2008). *Campus law enforcement, 2004–2005*. Bureau of Justice Studies Special Report: Washington, DC.

Romero, D. (2011, November 28). UCLA protesters to be in full effect today as UC Regents consider costs: Occupy principles invoked. *LA Weekly*. http://blogs.laweekly.com/informer/2011/11/ucla_tuition_protest_occupy_westwood_refund_california.php

Salter, C. (2011). Activism as terrorism: The green scare, radical environmentalism and governmentality. *Anarchist Developments in Cultural Studies*, *2*(1): 211–238.

Sclove, R. (1995). *Democracy & Technology*. New York: Guilford Press.

Simmons, L. (2010). The docile body in school space. In Torin Monahan and Rodolfo D. Torres (Eds.), *Schools Under surveillance: Cultures of control in public education*. New Brunswick, NJ: Rutgers University Press.

Sloan, J. J. (1992). Modern campus police: An analysis of their evolution, structure, and function. *American Journal of Police*, *11*(2): 85–104.

Sloan, J. J., Fisher, B. S., and Wilkins, D. L. (1996). Reducing perceived risk and fear of victimization on campus: A panel study of faculty members, staff, and students. *Journal of Contemporary Criminal Justice*, *12*(1): 81–107.

Stanfill, A. (2010, March 3). Sources: UF student was shot in head by police. *Gainesville Sun*. http://www.gainesville.com/article/20100303/ARTICLES/100309832?tc=ar

Starr, A., Fernandez, L. and Scholl, C. (2011). *Shutting down the streets: Social control and political violence in the global era*. New York: New York University Press.

Sum, C. (2010, March 25). According to chief, police not equipped in event of gunman. *Pasadena City College Courier*. http://www.pcccourier.com/news/according-to-chief-police-not-equipped-in-event-of-gunman-1.2422745#.T03Wjcx8-LM.

Virilio, P. (2006). *Speed and politics*. Los Angeles: Semiotext(e).

Virilio, P., and Lotringer, S. (2008). *Pure War*. Los Angeles: Semiotext(e).

Wada, J.C., Patten, R., and Candela, K. (2010). Betwixt and between: The perceived legitimacy of campus police. *Policing: An International Journal of Police Strategies & Management*, 33(1): 114–131.

Wilson, C. P., and Wilson, S. A. (2011). Perceived roles of campus law enforcement: A cognitive review of attitudes and beliefs of campus constituents. *Professional Issues in Criminal Justice*, 6(1–2): 29–40.

Wright, S. (1978). New police technologies: An exploration of the social implications and unforeseen impacts of some recent developments. *Journal of Peace Research*, *15*(5): 305–322.

Wyss Institute. (2011, November 17). Wyss Institute signs licensing agreement with K-Team Corporation. *Wyss Institute at Harvard*.http://wyss.harvard.edu/viewpressrelease/71/wyss-institute-signs-licensing-agreement-with-kteam-corporation-

• CHAPTER EIGHT •

We Are All Hokies

Surveillance Culture and Communication Technologies on a Post–Virginia Tech Campus

Caroline K. Kaltefleiter

> The Hokie Nation embraces our own and reaches out with open hearts and hands to those who offer their hearts and minds. We are strong, and brave, and innocent, and unafraid. We are better than we think and not quite what we want to be. We are alive to the imaginations and the possibilities. We will continue to invent the future through our blood and tears and through all our sadness.We are the Hokies.
>
> —*We Remember* by Nikki Giovanni, Virginia Tech Professor and Poet (2007)

The Virginia Tech tragedy of April 2007 constitutes a critical flashpoint in the development of emergency response systems at colleges and universities across the United States. More crucial though, the Virginia Tech incident marks the erosion of individual rights and a new culture of surveillance within the walls of the academy. In the wake of the Virginia Tech massacre, academic institutions responded with messages of solidarity and grief. A solitary campus mourning the loss of their own became a collective spectacle in which the nation's systems of higher education pledged to not only mourn the loss of its academic citizens, but also to take action by any means necessary to ensure that future such attacks be prevented, equating Virginia Tech to September 11th. In athletic stadiums, student union buildings, and on campus balconies, homemade banners read, "Today, We are All Hokies" (Educause, 2008, 5). The slogan resonated with university emergency managers, police leaders, information technology managers, and environmental officers. "We are All Hokies" connotes the idea that all campuses are vulnerable to attacks of violence, flu pandemics, and potential natural disasters. While the national media dissected the Virginia Tech tragedy, campuses across the country assembled their own emergency response teams to review campus procedures and technologies available to them.

This chapter examines the development of campus-wide alert systems within the framework of academic repression and surveillance in the wake of "campus terrorism" and "bio-terrorism." Theoretical analyses and references to Frederick Wiseman's films *High School* and *Titticut Follies* situate surveillance culture within filmic frames that articulate a continuum of watching and documenting the movements of those confined to institutions such as schools, universities, prisons, and asylums—with one's every move being recorded on security cameras, archived, and uploaded to a digital database. Attention is given to the use of new technologies such as automated phone calls, text messages, social network sites, data mining, and global positioning systems, used not only to disseminate information to campus constituents but also to gather and track information about its academic citizens. Finally, grassroots actions of resistance are explored to combat technological surveillance and lockdowns in the academy.

Flashpoints: 9/11 to Virginia Tech: Surveillance Culture and the Patriot Act

> So long as we do what they tell us, we have nothing to hide and nothing to fear.
> Derrick Jensen and George Draffan, *Welcome to the Machine* (2004, 133).

On the morning of April 16, 2007, campus residents of Virginia Polytechnic Institute and State University (Virginia Tech) awoke to a series of violent attacks when Seing-Hui Cho, a 23-year-old English major, opened fire and killed 33 people before taking his own life. Since then, communication technologies, namely email, cell phones, and even social networking sites such as Facebook and Twitter have been used to deliver relevant information needed to campus residents and to conduct surveillance, academic reconnaissance, if you will, on college campuses—all perfectly legal in the garrison state of post-9/11 America. In March 2008, the House of Representatives, in a closed session, the first since 1983, deliberated on updating electronic surveillance laws, focusing on a warrantless wiretapping program designed to monitor communications of U.S. citizens. On July 9, 2008, the United States Senate passed H.R. 6034, the Foreign Intelligence Surveillance Act Amendments Act of 2008. The new law modifies sections of the 1978 ("FISA") to expand (subject to certain new checks) the federal government's surveillance powers and retroactively immunize telecommunication companies that cooperate with the warrantless wiretapping program brought to light in 2005 (Kaltefleiter and Nagel, 2010). President George W. Bush signed it into law on July 10, 2008. President Obama extended provisions in the Patriot Act on May 27, 2011. In response, privacy advocates suggest that the new law has unduly and unaccountably expanded domestic spying powers. Secrecy surrounding the wiretapping debate contributes to an ongoing "Big Chill" effect propagated by the USA PATRIOT Act, an acronym for Uniting and Strengthening America by Providing Appropriate Tools Required to Intercept and Obstruct Terrorism. In a previous article on academic repression, my colleague Mechthild Nagel and I provided an in-depth analysis of the PATRIOT Act. As George Caffentzis notes, the PATRIOT Act defines "terrorist activity" so broadly that a police dragnet could easily entrap trade union picketers, raid the homes of animal-rights activists, or arrest people who donate to humanitarian organizations whose outreach include Palestinians or Iraqis. Here we examined two of the vague and ominous felonious acts described and proscribed in the PATRIOT Act:

1) "Terrorist activity": using a "weapon or dangerous device (other than for mere personal monetary gain)"; soliciting membership for, and providing material support to a "terrorist organization," even if this group is explicitly humanitarian.
2) "Terrorist organization": "two or more individuals, whether organized or not" who engage in terrorist activity.

This vague umbrella term certainly outdoes the infamous RICO Act of the 1970s, ostensibly directed against drug dealers and the U.S. Mafia, but more specifically used to round up leftist revolutionary dissenters, such as the Ohio 7 (Kaltefleiter and Nagel, 2010). Anti-terrorism laws have a long tradition in the U.S. In particular, beginning with the jailing of Eugene Debs for his unpatriotic dissent over World War I under the U.S. Sedition Act of 1918, which reads,

> Whoever, when the United States is at war, shall willfully make . . . false reports, . . . utter, print, write, or publish any disloyal, profane, scurrilous, or abusive language about the form of government of the United States . . . teach, defend . . . in favor of the cause of any country with which the United States is at war . . . shall be punished by a fine of not more than $10,000 or imprisonment for not more than twenty years, or both.

The USA PATRIOT Act granted law enforcement unprecedented powers of surveillance made possible by new technologies, cooperation of the major communication companies, and the public's fear of "terrorism" (Kaltefleiter and Nagel, 2010). Today surveillance is more than spying. Reg Whitaker (2000) defines surveillance as the collection and analysis of information for purpose of control undertaken by the dominant institutions of modern society: the central state, business corporations, mass-media entities, and academic establishments.

The Virginia Tech Tragedy in April 2007, followed by the February 2008 shootings at Northern Illinois University, constitutes a flashpoint where American colleges and universities are witnessing subtle acts of repression through the implementation of revised emergency response procedures in the wake of "campus terrorism." Many schools use new technology as means to contact an entire campus population swiftly such as aggressive use of email accounts, automatic calls to campus phones, and postings on the school homepages. Michael Lisi argues that the assessment of safety efforts and conditions are of paramount concern for today's colleges and universities, "from training to using emergency alert systems and mass notification speaker systems to assessing updating emergency response plans (2009, 8). At my own institution in upstate New York, university officials called for students, faculty, and staff to register their personal cell phone numbers with campus security as part of a comprehensive plan to execute an efficient crisis communication plan. In fact, the entire State University of New York system, comprised of some 64 campuses, is now networked with an emergency alert system (SUNY NY-Alert) with individual campuses being responsible for crafting their own emergency response protocol. Eric Gullickson, director of media relations at SUNY–New Paltz notes, "Virginia Tech was the catalyst that made every campus realize what its emergency alert response would need to be" (Lisi, 2009, 9).

The State University of New York entered into an agreement with the State Emergency Management Office (SEMO) to develop an emergency alert system. Students enter their numbers into the system for use in the case of emergency. The alert is sent to whatever contact information a person enters in the system (phone, email, cell phone, text, fax, etc.). Students are obligated to give their contact details as part of their registration process. Meanwhile, faculty have the "option" of disclosing their personal contact details and cell phone number, but are "highly encouraged" to have that information on record to be accessed should an emergency occur. While some assurances have been made to students and faculty that their stored personal information will be used for emergency purposes only, data collected, under the guise of protection, is another means by which the State exercises its ability to monitor its citizens.

Such databases could prove to be an invaluable tool to repress unions, activist, and community groups with the ability to access and to activate cell phones/numbers as tracking devices, giving the whereabouts of individuals and their virtual/physical presence, and therefore giving state officials information on the location of those attending campus demonstrations, speak outs, and acts of civil disobedience with precise accuracy. Faculty resistance to registering their cell phone is often met with a series of justifications, couched within a discourse of counter-terrorism, in an attempt to ensure everyone's safety/loyalty within the university. Those who resist, referring to civil liberty issues, may run the same risk as those who are deemed "unpatriotic" when calling for an end to U.S. military interventions. Here a sense of "unconcern" for terror threats or emergency situations is coded as disloyalty. Safety, not liberty, is of utmost concern— "you want to be safe, don't you?"Organizations such as SEMO capitalize on the public's/faculty's fear to create campus "safety," such as enforced through a 24/7 monitoring system at the disposal of the State (Kaltefleiter and Nagel, 2010, 410).

While campus cell phone registration may be considered an innocuous mode of surveillance taking place on campus today, a more striking and obvious visual form of repression is the recent construction of closed-circuit television systems and public-address systems mounted in and on university buildings. These mammoth structures create electronic gates and digital watch towers.In the fall of 2008, SUNY Cortland erected 22 public address speakers that resemble large megaphones, complete with a digital response signal that can be directed at five zones across our campus. The technology has the ability to create messages that can be targeted to a particular zone or all zones. Alert messages originate from the university police department. The presence of these speakers atop university buildings is akin to public-address systems used to direct prisoners at correctional facilities. Campus residents go about their normal day, perhaps forgetting that their movements may be monitored continuously. This sense of voyeurism embodies the notion that one's imprisonment is not in a physical building, but rather rests in a mental state of fear. In the end, surveillance then becomes redefined as the gathering of information and the supervision of subject populations in organizations such as schools, universities, prisons, or asylums (Dandeker, 1990).

Lights, Camera, Action in the Panopticon: From High School Hallways to University Corridors

Frederick Wiseman is a documentary filmmaker whose films about American institutions include *Titicut Follies* (1967), a film about a prison for the criminally insane, and *High School* (1968), where he documented the everyday life experiences of students, teachers, and administrators at North East High School in Philadelphia. The film reveals a militaristic, authoritarian environment. Wiseman's films have been subject to controversy, often criticizing societal structures and actions or inactions within institutions, including, asylums, prisons, and schools. The structure of Wiseman's work is as subtle and as complex as the institutions he examines. Cynthia Lucia writes, "The very acts of framing a shot and editing the final film shape our experience of each institution" (1994, 2). Of all of his films, *Titicut Follies* received the most criticism and was banned after the government of Massachusetts sought an injunction banning its release. The government claimed that the film violated the patients' privacy and dignity. Lucia (1994) describes *Titicut Follies* as a harrowing observational documentary about the Bridgewater State Hospital for the Criminally Insane that "reveals the dehumanizing treatment of inmates, stripped both literally and figuratively of their privacy, dignity, and identity" (1994, 1). As Lucia notes,

> Films such as *Titicut Follies* and *High School* do anything but reserve judgment—extreme choices in camera position alone, strongly reflect the filmmaker's response to each institution and people who operate within it. (2)

Wiseman's films are situated theoretically within the framework of cinema verité, sometimes called "observational cinema" in which the viewer experiences action as it happens. The lens of the camera creates its own form of a Panopticon, a reference to a prison structure designed by philosopher and social theorist Jeremy Bentham in 1785. Bentham's design incorporates a circular structure in which a guard tower with louvered windows is positioned in the center so as to create the illusion of being watched from all sides and angles. The concept of the design is to allow an observer to observe (-opticon) all.

Bentham's ideas extend beyond architectural significance and serve to provide a theoretical context to examine greater society. Jensen and Draffan (2004) write, "The Panopticon has become a model for the entire culture" (9). Michel Foucault, in his analyses of prisons and prison culture, noted that the Panopticon has become not only a simple idea in architecture, but also a metaphor for the power relations that undergird modern civilization. According to Foucault, "The major effect of the Panopticon is to induce in the inmate a state of conscious and permanent visibility that assures the automatic functioning of power. So to arrange things that the surveillance is permanent in its effects, even if it is discontinuous in its action; that the perfection of power should tend to render its actual exercise unnecessary. Jensen and Draffan note, "This architectural apparatus should be a machine for creating and sustaining a power independent of the person who exercises it; in short that the inmates should be caught up in a power situation of which they are themselves the bearers (2004, 10).

The semiotic images associated with the Panopticon connote representations of power. Bentham laid down the principle that power should be visible and unverifiable. These images are reified in the tall outline of a central guard tower from which he is being spied upon, creating a continuum of impressions of surveillance. To Bentham, the unverifiable is the notion that the inmate must never know whether he or she is being observed at any one moment; he or she must always live with the notion that the presence of a guard is always there. In order to make the presence or absence of the guard or inspector unverifiable, so that the prisoners in their cells cannot see a shadow, he provided architectural details inside/outside the Panopticon structure. Jensen and Draffan (2004) point out,

> Bentham envisaged not only Venetian blinds on the windows of the central observational hall, but on the inside, partitions that intersected the hall at the right angles and, in order to pass from one quarter to the other, not doors but zig-zag openings, for the slightest noise, a gleam of light, a brightness in a half-opened door would portray the presence of a the guardian. (Ibid.)

Michel Foucault would go on to suggest that the Panopticon is a machine for dissociating the seeing/being-seen dyad: in the peripheral; "one is totally seen, without ever seeing, in the central tower, one seeing everything without ever being seen" (1971, 201–202).

The concept of the Panopticon is reified through the myriad ways people are watched and placed under surveillance. Frederick Wiseman's early documentary films set the stage for a technologically mastered surveillance society that pays particular attention to the reactions, interactions, and distraction of youth as depicted in his film *High School* (1968). The linear trajectory of being behind the lens puts the viewer in the same position as the guard in the central tower so as to see everything through the lens without ever being seen by the subjects who are captured and contained within the frame. This transgression of recorded life through the lens allows the viewer to become a voyeur looking in on the experiences of those subjects in the frame and subsequently acts as a surrogate guard to keep control of the prisoners, or in this case, students. The opening-shot sequence of *High School* begins with a grainy black-and-white image of a delivery truck advertising "Penn Maid Products," an intertextual reference to the fact that many of the products students use in school such as notebooks, bluebooks, and desks are manufactured by prisoners. The camera follows the truck, and acts as a cinematic voyeur as the vehicle travels down a road parallel to a tall chain-link fence. Wiseman shoots through the links of the fence. A building comes into focus and connotes an image of incarceration. The truck, now in full focus, is a milk delivery truck, and it arrives within the gates of an institutional complex. On first glance, one expects it to be a prison or factory, but once the truck backs up to the loading dock, and begins to unload crates filled with pints of milk, the viewer recognizes the structure as a public school building, giving viewers privileged access to a cordoned-off structure known as North East High School.

Wiseman's films raise a number of questions not only related to the observational film debate, but also a contemporary exploration of voyeurism, privacy, surveillance, security, justice, and social change. Wiseman's work is significant to understanding the relationship between the prison-industrial complex, military-industrial complex, and that of the academic-industrial complex. Films such as *Basic Training* (Wiseman,1971) and *High School* (1968) weave a disquieting and disturbing tapestry of a woven narrative of incarceration, militarization, surveillance, and education. For instance, the detention scene in *High School* portrays students in a tiny concrete room—one that resembles a prison cell, or worse, a solitary confinement hole in the ground. The student attempts to plead his case with the dean of students, who during the exchange never looks up. The kid subsequently becomes exasperated. The dean issues a dictate: "You're suspended."

The opening shots in *High School* make it look like a factory or internment camp, yet it is only at the end when the school's principal reads a letter from a former pupil in Vietnam that the significance of the image becomes clear. The soldier is, he says, "only a body doing a job," and the school is a factory, for producing just such expendable bodies. Wiseman notes the interrelatedness of *High School* and *Basic Training*:

> At the time, many young men went right from high school into the army. *High School* was shot in 1968 and *Basic Training* was shot in 1970—the middle of the Vietnam War. They're both about education. In many ways the Army was more successful in its education methods than North East High School was. (quoted in Lucia, 1994, 7)

Segues between the films underscore educational indoctrination and a system that forgoes critical thinking and reflexivity. Wiseman notes,

> The army had a method, which had been tested and was working. They knew how to convert civilians into soldiers, how to train people who had no experience and no real interest in killing to become killers—in the service of the State—but killers nonetheless. They knew how to strip off that thin veneer of civility. (Ibid.)

In all, this trilogy of films by Frederick Wiseman offers insight into an understanding of the interrelatedness of the military-, prison-, and educational-industrial complex and just how important the freedom to think critically about information is to the developing mind. Wiseman's films present difficult complex subject matters without prescribing defined solutions or desirable outcomes. Meaning is generated in the viewers' interaction with the text, being an observer and confronting circumstances that are uncomfortable and uneasy to watch. In this cognitive space, the viewer may be moved to engage in a dialogue that will allow for critical pedagogy, direct action, and social justice to flourish so as to combat an age of surveillance culture where privacy and inalienable rights are risk, especially for the youngest members of American society.

Radio Frequency Identification: Tracking Kids and Targeting Teens

Deprogramming contemporary society relies on reprogramming the youngest minds not to question advancements in technology that allow for the gathering of information and supervision of students/citizens reified in Radio Frequency Information tags (RFIDs). Katherine Albrecht (2002) articulates the connections between auto-ID technology and its impact on consumer culture and how it is positioned to take over our lives. Jensen and Draffan note, "Auto-ID couples with radio frequency (RF) identification with highly miniaturized computers that enable products to be identified and tracked at any point along the supply chain (2004, 5). These tags are cost efficient, less than one cent, and may be incredibly small—the size of a grain of sand or a speck of dust. "They are already being built into clothing, auto parts, drugs, and some foods. Receiver and reader devices are used to pick up the signal transmitted by the RFID tag" (2004, 5). In *The Filter Bubble*, Eli Pariser advises, "suddenly it's possible for business to track each individual object they make across the globe" (2011, 197). Pariser notes that younger demographics are targeted for RFID use. He cites the Coca-Cola Village Amusement Park in Israel. In 2010, Facebook teamed up with Coke, two brands well known to the 12- to 24-year-old demographic. According to Pariser,

> Teenagers who attended the park were given bracelets containing a tiny piece of circuitry that allowed them to "Like" real-world object. Wave the bracelet at the entrance to a ride for example and a status updated posted to your account testifies that you're about to embark. Take a picture of your friends with a special camera and wave the bracelet at it and the photo's posted with your identity already tagged. (197)

The bracelets are all embedded with RFID chips. RFIDs use call-and-response technology, allowing for items and people to be tracked. The chips turn out a unifying code that, when correlated with a Facebook account, grants information of not only the wearer of the bracelet, but also individuals with whom the bracelet wearer interacts. Such integrated marketing and product placement is a media extension of myriad advertising efforts towards children, young kids, and teens that gained momentum with Channel One television. Schools who adopt Channel One programming are required to screen advertisements to their pupils each day. Mary Story and Simone French (2004) note that "[a]bout 12,000 schools or about 38% of middle and high schools in the US are connected to Channel One, the 12-minute current events program that carries two minutes of commercials including advertisements for soft drinks and high fat snack foods (2004, 4). Schools receive free video equipment in exchange for mandatory showing of the program in classrooms. Brand and Greenberg (1994) evaluated the effects of Channel One in-school advertising on high school students' purchasing attitudes, intentions, and behaviors. About 70 percent of the 45 food commercials shown on Channel One during one month were for food products including fast foods, soft drinks, chips, and candy. "In schools where Channel One was viewed, students had more positive attitudes about the advertised products, and were more likely to report intentions to purchase these products compared to students who did not have Channel One in their classrooms" (Story and French, 2004, 4). The connection between consumers and students may begin with entertainment, but extends far beyond the social environment of younger children and teenagers. A day at the amusement park or watching television is seen as fun and not intrusive and serves to acclimate kids to conditions of surveillance in schools. The usage of RFID chips continues to generate debate in school districts across the United States. In California, school superintendents struck a deal with a local RFID maker as a beta tester to track attendance and weed out trespassers. Kim Zetter of *Wired Magazine* describes how the system would be used at Brittan Elementary School in California. According to Zetter (2005):

> The system consists of a photo ID card affixed to a lanyard and worn around the neck. Embedded in the card is an RFID chip that contains a 15-digit number assigned to each student. As students pass beneath a doorway scanner on their way into a classroom, the scanner records the number and sends it to a server in the school's administrative office. The server translates the digits into names and sends an attendance list to the teacher's PDA, identifying all of the students who walked through the door. The teacher then visually verifies that the names on the PDA list match the students in the classroom.

Brittan was the first school in California to use RFID, but not the first in the nation. Zetter (2005) reports, "Spring Independent School District near Houston, Texas, recently gave 28,000 students RFID badges to record when students get on and off school buses. School administrators monitor the information to prevent child abductions and truancy." In addition to tracking students in classrooms, some principals want students tracked in bathrooms and locker rooms. Teachers become de facto police or prison guards watching and reporting movements of their students to a larger system or "Big Brother" akin to what George Orwell wrote about in his book *1984*.

Not surprising, the American Civil Liberties Union and parents have raised privacy and civil liberties concerns with regards to RFIDs in schools. Some students questioned these practices as well. One pupil wrote,

> RFID chips on school uniforms are incredibly invasive, and it is very easily foiled to boot. I am a student myself and I would take this to be a serious insult on my maturity and judgment. What are they going to do next, install

microphone surveillance bugs on our foreheads? (CCTV Surveillance Systems, 2011)

One cannot overlook financial deals made at the expense of students and their privacy. Marc Perton (2005) notes "parents questioned the financial relationship between the school and the cards' manufacturer, InCom, the company that designed the badge system and is paying the school to try out the cards and offering it commission on future sales to other schools." The InCom business plan resembled that of Channel One in which the product/service is made available for free in exchange for access to a captive audience or demographic, in this case the students, who are like prisoners in some penitentiaries used as guinea pigs for experimental medical/social research.

In the end, the Brittan Elementary School abandoned their experimental RFID program after InCom pulled out of its agreement with the school due to mounting pressure and protests from parents, community members, and public advocacy groups (Parents and Civil Liberties Groups, 2005). In a co-authored letter, the Electronic Privacy Information Center (EPIC), along with the Electronic Frontier Foundation and ACLU-Northern California, urged the Brittan School Board to terminate the program that used mandatory ID badges to track children's movements in and around the school with RFID technology. The letter argued that the program breached children's right to privacy and dignity by treating them like cattle or pieces of inventory (Ibid.). While Brittan Elementary may have disbanded its RFID program, other schools have not. Every day, children in school districts across the United States don their RFID badges, board buses, and enter through the doors of the schools—scanned for their identity and authentication. Images of children wearing their RFID tags, riding on buses, and entering campus grounds resemble scenes from Frederick Wiseman's *High School* in which the buses and milk truck are tracked by the camera—entering the chain-link compound to deposit products and pupils (prisoners). In contemporary society, students become a captive audience or experimental population to test tracking technology and to familiarize these future citizens of America to conditions of surveillance.

Remote Cameras and College Campuses

Tracking technology is closely linked to surveillance cameras on college and university campuses. Frederick Wiseman's observational camera techniques and sense of omniscient presence in and out of the frame creates a theoretical foundation by which to examine security cameras and their role within the walls of the academy. Cameras allow university officials to document, watch, and scrutinize the everyday life actions of campus workers and residents. In his essay, "Smile! You're on Campus Camera," Jeffrey Young (2003) discusses camera surveillance on university campuses. Young notes,

> A growing number of colleges are setting up camera networks and expanding their camera systems. At the University of Pennsylvania, campus security workers are assigned to do "video patrols" of the campus, meaning that they cycle through live views from the cameras on a bank of monitors in the security control room, or at monitors on their desks, looking for signs of trouble. (2003, 2)

In some cases, university administrators and police officers attribute the necessity for more surveillance cameras to a rise in campus crime, including burglary, aggravated assault, attempted rape, and even murder. While security cameras may be used to deter incidents from happening, cameras have also been used to solve crimes. The University of Wisconsin–Superior, a small comprehensive public institution, has added a number of surveillance cameras across the campus over the past two years. Mike Bodin, a security officer at UW–Superior, recalled a series of car break-ins about two years ago at the university. Campus safety used surveillance footage of campus parking lots to help find the thief (Berrens, 2009).

Many elite universities in the United States have high-end surveillance systems that have been used to solve crimes ranging from petty larceny to murder. At Yale University, security cameras played an integral role in solving the disappearance of 24-year-old doctoral student Annie Le on September 8, 2009. Security video revealed Le entering the building. When Le went missing, police discovered by observing security camera footage, she had never actually left the research building in which her body was found. The medical research complex houses more than 75 security cameras. Le's body was found within the walls of the medical research building where she worked. Raymond Clark III, a Yale co-worker at the research facility, was convicted of Le's murder and was sentenced to 44 years in prison (Gideon, 2011).

While cases such as the murder of Annie Le certainly caused alarm on many university campuses and raised security measures, potential attacks of animal research laboratories by animal-rights groups has also led to more stringent security at research facilities. In fact, the type of electronic equipment used in the medical complex at Yale has increasingly become part of the regular framework of animal research labs. The heavy security measures are designed to protect people and property as well as the animals that provide valuable information for researchers conducting scientific experiments. There are more than 1,000 research facilities in the U.S. registered with the Agriculture Department, and many are tied to universities. Richard Bianco, the head of experimental surgery at the University of Minnesota, says it's a challenge research facilities have had to address. His school began to do so a decade ago after animal-rights activists came on campus (Corley, 2009). According to Bianco, "The university put cameras in hallways and common areas, required key cards for access into certain lab rooms and facilities, and put up panic alarms—buttons that could be pressed to alert police to come to the area" (Ibid.). The ratcheting up of surveillance technology on university campuses, particularly around animal research labs, resembles images of science-fiction films—only this is reality. Jack Hessler, the co-editor of a book on planning and designing research facilities—a project of the American College of Laboratory Animal Medicine—says that beyond setting up cameras, card swipes, and key card controls, some facilities use biometrics like a handprint or retina identification (Hessler and Lehner, 2008).

Surveillance cameras are not always used for security measures or so it would seem. In fact, just as the tracking bracelets in the Coca-Cola Amusement Park helped introduce teens to RFID badges via entertainment, so too have webcams become incorporated in the landscape of colleges and universities, providing viewing entertainment and information. Some schools have a web camera directed at campus monuments and structures 24 hours a day, seven days a week. At my own campus, a webcam records weather conditions at our experiential learning center in the Adirondacks. One can watch the change of seasons and people at the camp station via the webcam. On our main campus, a webcam system was installed throughout all of our dining halls this year. Auxiliary Services Corporation oversees all the dining halls. ASC issued a statement proclaiming that the new system would provide greater efficiency and convenience for students. In a statement about the new webcam system, ASC noted that inspiration for this project came from a computer-science laboratory project in England.

> In 1991, researchers in a University of Cambridge computer laboratory searched for a solution to a very practical question—was there coffee in the Trojan Room, space within the computer lab. The coffee pot was shared between 15 researchers and a freshly brewed pot rarely lasted long. To address this, researchers pointed a video camera at the pot, wired it to their network, and created the world's first Web camera. (Auxiliary Services Corporation, 2010)

In the spirit of the Trojan room and Cambridge webcam, the ASC statement noted, "For student convenience, SUNY Cortland's new webcam system will now give students live-stream access to SUNY Cortland dining facilities, Dunkin Donuts, Dragon's Den, Neubig and Hilltop" (Ibid.). Shortly after this announcement, students in my Issues and Digital Culture class began talking about the notion of peer-to-peer monitoring, the new webcam system, and their uneasiness about it. In one class, we brought the system up on the big screen in the lecture hall. We logged on to the Hilltop dining hall and watched fellow students and faculty select, wait, and pay for their food. We watched the expressions of those on camera—evaluating their stature, wardrobes, and even whereabouts as to why they were not in class. One student on camera was supposed to be in my lecture as he stood in line waiting for his deli sandwich with his buddies. Students giggled to alleviate the discomfort they felt while "spying on their friends."

In a separate instance, one student, Brenna Filipello, wrote about discovering the reality of the campus webcams and the feeling of being put under the microscope. She relates having a conversation with a friend during breakfast at the dining hall.

> I said to my friend—Why are you here? I thought you had a class at this time.
>
> I do. I'm actually leaving right now. I was just checking my email before class and I stumbled across the link for SUNY Cortland's webcam system.
>
> I watched you sitting alone in the dining hall from my BlackBerry and I thought I'd visit you!
>
> Without any further explanation, she got up and left. For the rest of my breakfast, I couldn't shake the feeling that someone was watching me. (Filipello, 2010)

The dilemma of peers watching one another on webcams made national headlines with the Tyler Clementi case, a Rutgers freshman who committed suicide after finding out his roommate had remotely activated a webcam in their dorm room and watched Clementi having sex with a man. The webcam feed went beyond the Rutgers campus and the perpetrator of the crime, Dharun Ravi, used social networking sites and invited others to watch Clementi's tryst as it unfolded live on camera. Ravi was indicted for invasion of privacy, bias, and tampering with evidence in Middlesex County, New Jersey (Ali, 2011). He was convicted of 15 criminal counts, including bias intimidation and invasion of privacy and would go on to serve 20 days out of a 30-day sentence for his crimes.

The Clementi case exemplifies links between social voyeurism and power. Anders Albrechtsund (2008) extends Foucault's notion of power relations of the Panopticon to peer-to-peer monitoring and argues that these technologies actually amplify the top-down monitoring. He notes, "As a consequence, law enforcement technologies are brought into social life." Mark Andrejevic discusses the notion of peer surveillance in modern society

> The participatory injunction of the interactive revolution extends monitoring techniques from the cloistered offices of the Pentagon to the everyday spaces of our homes and offices, from law enforcement and espionage to dating, parenting, and social life. In an era in which everyone is to be considered potentially suspect, we are invited to become spies—for our own good. (2005, 494)

Remote cameras and webcams epitomize surveillance culture on colleges and universities today. The continuum of being watched by others or watching one's peers in a mundane setting such as a dining hall serves as a mechanism by which to condition students to accept being monitored 24/7. This behavioral conditioning extends beyond their college years and readies them to a culture of surveillance that records their everyday life experiences; be it under the guise of safety, fear of terrorism, or convenience to get their daily coffee or fast-food meal. As such, private lives become public ones and the contemporary public morphs into one entity. Citizens engage in participatory surveillance, making us all spies in a disciplinary society.

Critical Reflecion, Direct Action, and Self-Liberation

Zygmunt Bauman, author of *Modernity and the Holocaust*, notes, "The terror by which the Panopticon is maintained only remains effective as long as the balloon of rationality is pricked" (1992, 203). In an age of electronic surveillance, there is a need for critical reconnaissance in an attempt to expose surveillance actions taken inside/outside of the academy that constitute repression. The work of Marxist revolutionary and political prisoner Ché Guevara lays the foundation for critical reflection. Guevara spoke of the necessity to create critical forums of education and cultural resistance until his death in 1967—a murderous act commissioned by the United States Central Intelligence Agency (CIA). According to Guevara, "[s]tudy . . . and the place of study where you carry out your work is the patrimony of no one—It belongs to all the people . . . and it must be extended to the people or the people will seize it" (quoted in Retamar, 1971, 71). Guevara's words call to mind the work of Antonio Gramsci, an Italian Marxist intellectual who was sentenced to more than 20 years in prison for his critique of institutions used by the State to repress its citizens. Gramsci's work illuminates the idea that "the machine" does not exist, but rather it is a mere suggestion, a device like the Panopticon that can alter the behavior of those in the system.

As Jensen and Draffan point out, the machine "like other social constructs has consequences, . . . but it does not really exist. We pretend it does, but our pretensions do not making it real" (2004, 223). We must be engaged citizens so as to think about the realities of the situation. For Gramsci, the Italian prosecutor at his sentencing, urged the court to pronounce a sentence of civil death: "For twenty years, we must stop this brain from working" (Forgacs, 1971, 22). This imploration haunts the halls of the university today, silencing critical discourse through personnel management and citizen surveillance.

To combat future civil deaths and citizen surveillance, we must explore a dialogue of resistance through the use of direct action, collaborative participation, and evaluation of social, cultural, historical, and political artifacts such as Frederick Wiseman's films *High School, Titicut Follies, and Basic Training.* Wiseman's work allows for critical reflection and raises important questions related to our new surveillance society: "at what point in the act of bearing witness does the camera merge with the institution and further strip its subjects of their dignity?" (Lucia, 1994, 1). Such questions speak to a devolving notion of identity wherein one's sense of being is becoming eclipsed by filmic frames, genetically modified organisms, manipulated images, readable code, or encrypted numbers. As Jensen and Draffan note, "How much easier we make it for those at the center of the Panopticon if we allow them to rob us of our complex and willful identities" (2004, 56).The observational lens flows into the security lens framing those inside and outside the frame.

As I was completing this essay, the Federal Bureau of Investigation announced plans to issue an updated version of its manual known as the *Domestic Operation and Investigations Guide*—a document that outlines parameters for investigating individuals. According to the *New York Times*,

> New guidelines will allow FBI agents to investigate people and organizations "proactively" without firm evidence for suspecting criminal activity. The new rules will free up agents to infiltrate organizations, search household trash, use surveillance teams, search databases, conduct lie detector tests, even without suspicion of any wrong doing. (Savage, 2011)

Civil liberties advocates continue to raise alarm over news that the FBI is giving agents more leeway to conduct domestic surveillance. Colleges and universities are exempt from these new rules. As a critical theorist, peace educator, and social activist, I am particularly concerned about the new rules allowing FBI agents to infiltrate activist organizations, to expand spy procedures, and to target resistance groups. In my own work, I argue for the necessity of critical pedagogy, cultural resistance, and political activism in the wake of recent assaults on academic freedom, civil liberties, and privacy by exposing the terrorism myth and its role in the (re)production, distribution, and consumption of knowledge in a capitalist society. In a collective response, scholars/activists must confront academic repression and civil liberties head on by advancing competing global dialogues that emphasize justice, democracy, and liberation. To act collectively with one voice is more important than ever today, since individual faculty protesting repression can easily be silenced, if not dismissed.

Physical and digital connections alike are crucial to creating a network of critical recognizance whereby the university is watched by those on the inside/outside and held accountable for punishments imparted to those who speak out against the corporatization of the university. As Jensen and Draffan note, "Until those in power figure out a way to completely replace humans with machines. . . there will always be more of us than there are of them"

(2004, 225). In this sense, we truly are all Hokies. We are willing to contest manufactured realities; to take action against academe's complicity in creating a just world; and to draw back the curtain of the machine, exposing the filmic frames of the wizard and his Panopticon.

Bibliography

Albrecht, K. (2002, Summer). Auto-ID, tracking everything, everywhere: Supermarket cards, tip of the retail surveillance iceberg. *Denver University Law Review, 79*(4).

Albrechtsund (2008). Online social networking as participatory surveillance. *First Monday, 133*(3). http://www.uic.edu/htbin/cgiwrap/bin/ojs/index.php/fm/article/view/2142/1949

Ali, A. (2011, April 20). Roommate indicted in Rutgers student suicide. *Reuters.* http://www.reuters.com/article/2011/04/20/us-rutgers-suicide-indictment-idUSTRE73J4NR20110420

Andrejevic, M. (2005). The work of watching one another: Lateral surveillance, risk, and governance.*Surveillance & Society,* 2(4), 479–497. http://www.surveillance-and-society.org/articles2(4)/lateral.pdf

Auxiliary Services Corporation. (2010). http://www.cortlandasc.com/neubig-dining-webcam.cfm

Bauman, Z. (1992). *Modernity and the Holocaust.* Ithaca, NY: Cornell University Press.

Bentham, J. (1995). *The panopticon writings.* London: Verso.

Berrens, B. (2009, September 25). Smile, you're on camera. *The Stringer.*http://www.uwsuper-stinger.com/smile-youre-on-camera.html

Brand J. E., and Greenberg, S. (1994). Commercials in the classroom: The impact of Channel One advertising. *Journal of Advertising Research,* 34:18–27.

CCTV Surveillance Systems. (2011, May 27). What is your opinion on the recent use of RFID trackers in school systems? http://www.usedbarcodescanner.com/forum/what-is-your-opinion-on-the-recent-use-of-rfid-trackers-in-school-systems/

Corley, C. (2009, September 17). Yale killing defied tight lab security. National Public Radio. http://www.npr.org/templates/story/story.php?storyId=112933564

Dandeker, C. (1990). *Surveillance, power, and modernity: Bureaucracy and discipline from 1700 to the present day.* Cambridge, U.K.: Polity Press.

Educause. (2008, October). The role of IT in campus security and emergency management: White paper.http://net.educause.edu/ir/library/pdf/PUB9001.pdf

Filipello, B. (2010). The lifestyle of the visually impaired. Unpublished paper. SUNY Cortland.

Forgacs, D. (1988). *An Antonio Gramsci reader.* New York: Schocken Books.

Foucault, M. (1971). Discipline and punish: The birth of the prison. New York: Vintage Books.

Gabbard, D. (2008). *Knowledge and power in the global economy: The effects of school reform in a neoliberal/neoconservative age.* Mahwah, NJ: Lawrence Erlbaum.

Gideon, G. (2011, June 3). Raymond Clark III sentenced to 44 Years. *Yale Daily News.*http://www.yaledailynews.com/news/2011/jun/03/raymond-clark-iii-sentenced-44-years/

Giovanni, N. (2007, April 17). *We Remember.* Transcript of Convocation Address to commemorate Virginia Tech Tragedy. http://www.remembrance.vt.edu/2007/archive/giovanni_transcript.html

Hessler, J., and Lehner, N. (2008). *Planning and designing research animal facilities.* London: Academic Publishers.

Jensen D., and Draffan, G. (2004). *Welcome to the machine: Science, surveillance, and the culture of control.* White River Junction,VT: Chelsea Green Publishing.

Kaltefleiter, C., and Nagel, M. (2010). The carceral society: From the prison tower to the ivory tower. In A. Nocella, S. Best, and P. McLaren (Eds.), *Academic Repression.* Oakland, CA: AK Press.

Lisi, M. (2009, October). A safe place to work and learn: UUPers work to keep the SUNY community free from harm. *Voice Magazine*, 37(2), 8–10.

Lucia, C. (1994, October). Revisiting "High School": An interview with Frederick Wiseman. *Cinéaste*, 20(4), 5–11.

Parents and Civil Liberties Groups. (2005, February 7). American Civil Liberties Union. http://www.aclu.org/technology-and-liberty/parents-and-civil-liberties- groups-urge-northern-california-school-district-t

Pariser, E. (2011). *The filter bubble.* New York: Penguin Press.

Perton, M. (2005, February 12). California school mandates RFID tags for students. *Engadget.* http://www.engadget.com/2005/02/12/california-school-mandates-rfid-tags-for-students/

Retamar, R. F. (1971). Caliban: Notes towards a discussion of culture in our America. *Massachusetts Review*, 15(1&2): 7–72.

Savage, C. (2011). F.B.I. agents get leeway to push privacy bounds. *New York Times.* http://www.nytimes.com/2011/06/13/us/13fbi.html

Story M., and French, S. (2004). Food advertising and marketing directed at children and adolescents in the US. *International Journal of Behavioral Nutrition and Physical Activity*, 1(3): 1–17. http://www.ncbi.nlm.nih.gov/pmc/articles/PMC416565/

Whitaker, R. (2000). *The end of privacy: How total surveillance is becoming a reality.* New York: The New Press.

Wiseman, F. (1967). (Director). *Titicut follies* .Cambridge, MA: Zipporah Films.

——. (1968). (Director). *High school.* Cambridge, MA: Zipporah Films.

——. (1971). (Director). *Basic training.* Cambridge, MA: Zipporah Films.

Young, J. (2003, June 13). Smile, you're on campus camera. *Chronicle of Higher Education*, 49 (40). http://chronicle.com/article/Smile-You-re-on-Campus/30029

Zetter, K. (2005, February 10). School RFID plan gets an F. *Wired Magazine.* http://www.wired.com/politics/security/news/2005/02/66554

• CHAPTER NINE •

Research, Repression, and Resistance

Scott DeMuth and David Naguib Pellow

Social Movements: Sites of Theory and Knowledge Production

Angana Chatterji is a scholar-activist and colleague, once told us that sustainable knowledge is a foundation of sustainable communities. That is to say, although some knowledge is produced for profit or personal gain, sustainable knowledge is produced in order to nurture all members of society and the ecosystems upon which we depend. This requires building relationships and institutions that are committed to democracy as well as social and environmental justice.

Many facets of society are involved in producing knowledge, but the academy has long been privileged as the main source of knowledge production and theory creation. All people engage knowledge on a daily basis; however, scholars are paid to do this through our teaching, research, and writing.

Unfortunately, our experiences as both practitioners and researchers within social movements have created dissonance, challenges, and questions regarding the roles as activists and scholars. Many tensions are unresolved and many questions continue to arise, but we see our roles as united in the pursuit of producing, testing, and transmitting theories and knowledge about social change.

The Highlander Folk School is an iconic example of this process within the context of social movements. The school was a critical site of movement training and learning for generations of activists, including participants in the civil rights, immigrant rights, and environmental justice movements. Activists such as Rosa Parks and Martin Luther King, Jr., shared and tested theories of and knowledge about social change tactics, strategies, and philosophies at Highlander, resulting in a much more robust movement that achieved significant social change. Scholars have also identified the role of Southern black churches and colleges in creating "free spaces," sites relatively free from surveillance that provided networks of potential recruits and space for emerging movements to develop oppositional identities, strategies, and tactics (Polletta 1999). These were also sites of tactical innovation, which led to the explosion of sit-ins throughout the 1960s (Morris 2009).

There are many other examples of these sites of innovation and transmission, which are crucial evidence that movements do more than simply inculcate agendas and mobilize people to action. They are spaces where knowledge and theories are debated, revised, rethought, and applied. This is a vital, yet often overlooked, component of what movements do.

In a critically important book on social movements, sociologists Ron Eyerman and Andrew Jamison make the point that

> A social movement is not one organization or one particular special interest group. It is more like a cognitive territory, a new conceptual space that is filled by a dynamic interaction between different groups and organizations. ...It is precisely in the creation, articulation, formulation of new thoughts and ideas—new knowledge—that a social movement defines itself in society. (Eyerman and Jamison 1991: 55).

Through our own experience as observers and researchers in social movements we have come to see them as important sites and forces for the creation and application of new and transformative knowledge. In addition, through our experience as movement participants we see social movements as experiments in producing social change. We believe social movements are important spaces in which learning, innovation, and knowledge production and dissemination take place. In that sense, one might think of movements and activists as practitioners of a much-needed science of social change.

Thanks to our engagement with social movements and Social Movement Theory literature, we believe social movements counter the hegemonic notion of the academy as the sole or central site of knowledge production. In addition to engaging in historical research regarding past movements, social movements make theoretical claims (explicitly or implicitly) and test these theories by both attempting to replicate results and developing innovative tactical approaches to social change.

Social Movement Theory literature is filled with cases, experiments, knowledge, and theories that have often been produced outside of, yet gleaned and appropriated by, the academy. In fact, these ideas are often extracted from movements and then published and copyrighted by academics whose careers are then made without any of the hard, mundane, and oftentimes risky work that is needed to produce this knowledge. This activity results in publications, promotions, and successful careers for such scholars, while the participants in social movements have been and continue to be at risk of being arrested, assaulted, or assassinated.

The literature of Social Movement Theory is not alone in this appropriation of knowledge. Whole disciplines have their origins in mapping physical and cultural territory and facilitating invasion, colonization, and genocide. Additionally, scholars have conducted research on behalf of the State and corporations in order to penetrate potential sources of cheap labor, consumer markets, and ecological wealth. Biologists, medical researchers, and other scientists have used public funds to appropriate Indigenous and folk knowledge for medicines that are then patented and privatized for profit.

Although these examples are rather easy targets for criticism and critique, there has been little analysis of how the more "progressive" elements of the academy engage in similar processes of appropriation and subjugation. Even though many scholars, especially those in the social sciences and humanities, conceive of themselves as producers and transmitters of theory, art, culture, and knowledge, the academy often resembles an extractive industry.

Repression as Research/Research as Repression

Similar to extractive industries, the academy colonizes cognitive, intellectual, and even physical territories in addition to its extraction and appropriation of knowledge. Karl Marx and Friedrich Engels remind us that "the ideas of the ruling class are, in every age, the ruling ideas: i.e., the class which is the dominant material force in society is at the same time its dominant intellectual force" (Marx and Engels 1998: 39). The academy should be suspect precisely because elites seek to shape and control the contours and content of ideas in public circulation.

Participants in social movements are often aware of the work of Social Movement researchers, and these researchers are often actually working in these movements. Theory is drawn from the experiences of movements, and the research of scholars, most often written for an academic audience, also finds its way into the discourses and strategies of social movements. One example of this would be how the concept and understanding of framing have been transmitted among social movements, to Social Movement Theory, then back into social movement organizing. For scholar-activists, this is often a central goal of research: we often hope to pick up on central processes within these movements but would also like to provide specific theories, tactics, or tools that will be adopted and guide future action.

There has been little discussion or analysis about how academic knowledge and theory may be privileged over the knowledge and theory produced by movements or how academic knowledge may serve to constrain and limit the intellectual terrain of movements and the repertoire of actions available. A prominent example is Social Movement Theory's emphasis on the role of the State in implementing social change and the need for movements to interact with and focus on the State. When academic knowledge about this process is privileged, this can implicitly support reformist movements, institutional incorporation of movement organizations into the political process, and the cooptation of movements.

Another unfortunate example is illustrated by some of the literature regarding anti-apartheid movements in South Africa. Much of the Social Movement literature during those mobilizations downplayed, dismissed, or even outright ignored the more radical, militant elements of the anti-apartheid movement. We can probably attribute this to the fear of taking unpopular stances with regard to forms of resistance and liberation that are considered too threatening. It was only years later that some academics began to affirm the role that armed struggle and underground networks played in producing success (Seidman 2001).

Similar dynamics have emerged within the academy in response to the current mobilizations to boycott and divest from Israel in response to the Israeli government's occupation of Palestinian land. The Palestinian Campaign for the Academic and Cultural Boycott of Israel (PACBI) was one of the founding entities of these campaigns and remains a key part of the Palestinian-led global Boycott, Divestment, and Sanctions (BDS) movement. The logic of the campaign is not only to implement a boycott to pressure the government of Israel to comply with international law and human rights law; it is also an acknowledgment that cultural and academic institutions play an important role in supporting or challenging institutional forms of discrimination. Many academics have signed on to the BDS campaign, but widespread fear and anxiety prevents countless others from doing so, supported by the highly visible firings and disciplinary actions taken against a number of scholars for speaking out in support of Palestinian rights. Fear is a powerful antidote to social change and scholars must work together to challenge it.

The cognitive repression and constraints of Social Movement Theory may be a little vague, but we also see clear resemblances between the research repertoires of Social Movement scholars and the repertoires of repression used by the State.

Here, we view the repression of social movements as more than just overt acts of brute force but rather as a *science*. In the *New State Repression* (2006), Ken Lawrence discusses how political repression has fundamentally changed from its past practice. Of course, there is certainly a difference in technologies used, but the biggest change is one of strategy and praxis. In earlier eras, rebellions and insurgencies were seen as pathologies and erratic occurrences that were put down with brute force. Today insurgency is seen as a near-constant threat that has to be constantly maintained through investigation and surveillance. This permanent state of repression is the full-time task of security forces.

The science behind repression is best defined in General Frank Kitson's *Low Intensity Operations: Subversion, Insurgency, and Peacekeeping* (2011). Repression begins not during rebellion or even during the active phases of social movements, but at the earliest stages of movement formation and mobilization. Kitson emphasizes the need to avoid the use of brute force in suppressing movements and instead focuses on intelligence gathering. This is crucial to the State's efforts, as the early stages of movement formation may provide the best opportunity to identify participants and infiltrate organizations.

Repression, like many forms of basic or applied science, is rooted in a desire to *know* and to develop and deploy knowledge for the advancement of particular interests (in this case, the interests of the State and capital). This knowledge results from careful, empirical observation, experimentation, data gathering, and analysis. Studies of institutional repression reveal quite clearly that State and corporate institutions essentially follow these protocols of basic science in order to protect their interests against threats and competitors (Boykoff 2007). In that sense, the science of institutional repression follows Foucault's reasoning that knowledge of a population leads to power and control over that group (1995). This can be seen in the specific cases where State surveillance and data gathering on activist organizations are used to quash or manage political dissent. For those groups deemed threats to the state, the science of repression can have serious consequences.

The U.S. government has been able to deport and physically exclude many dissidents since the nation's founding, and it has developed similar methods for domestic/citizen dissidents as well, including surveillance, harassment, firings, arrest, criminalization, deportation, and assassination. Following from our view of State repression as a science of domination through knowledge production, we assert that the knowledge the State *extracts* from movements is used to manage, contain, and control dissident bodies within current and future social movements. That is, such knowledge is used to exert influence over *potentially* dissident bodies—i.e., the public—to prevent the future development and growth of such movements. Again, although brute force and violence (or more commonly the threat of their application) are the underlying power of the State, institutional control, repression, and counter-insurgency occur through more subtle repressions of surveillance, suppression of ideas, and the shaping of public discourse (Boykoff 2007).

Through the above discussion of repression, we can see how academic research mimics some aspects of State repression. Even progressive scholars and scholar-activists run the risk of replicating aspects of the science of repression by participating in the everyday mundane tasks that comprise research. For example, scholars involved in fieldwork and ethnography within social movement settings collect data through interviews, focus groups, empirical observation, and participation in their research sites—all of which are routine components of the research process. Both the State and scholars extract knowledge from and about movements out of a desire to *know*. The repertoires of surveillance, infiltration, participation-observation, and informants are used simultaneously by the State and by scholars. State security forces, often through local police, campus security, or the FBI, attempt to observe, interview, and infiltrate movements. Participants often condemn the State for employing these repertories of repression against their movements, yet scholars are often surprised when the same condemnation is directed at them. Although there are certainly some key differences in the motivations behind State surveillance and scholarly surveillance—primarily, neutralization and disruption as opposed to support and replication—there is considerable overlap in the practiced repertories of repression and repertories of research.

In "Can There Be a Feminist Ethnography?", Stacey (1988) raises related questions regarding ethnographic methods, stating, "I find myself wondering whether the appearance of greater respect for and equality with research subjects in the ethnographic approach masks a deeper, more dangerous form of exploitation." Stacey (1988) is critical of the ethnographic process as being much more exploitive than it is often portrayed by social scientists. In fact, the ethnographic process may put subjects at more of a risk of exploitation than traditional positivist methods. Stacey focuses on the exploitation in the ways that researchers benefit from the relationships in the field, coupled with the researcher's ability to just pick up and leave, but her critique could be extended to the role this research plays a role in the surveillance and repression of the movements studied.

Once scholars publish their work, virtually anyone can access it, including and especially State authorities. Social Movement Theory broadly focuses on why social movements emerge, what kind of people join and support them and why, what their goals are, and how they work to achieve those goals. There is every reason to believe that State institutions have a strong interest in the Social Movement literature, as it potentially provides important insights into ways that these organizations work and also how they can be more effectively contained and managed.

Scholars run the risk of doing the work of State security forces if they become an outsourced, unofficial extension of the State's disciplinary apparatus. We must be conscious of this possibility. The data gathered on movements may be used against them, whether it is gathered by scholars or the State. But we see the dissemination of knowledge as a crucial difference between scholars and the State. The creation and dissemination of knowledge are crucial to building effective social movements. Scholarly research is often disseminated in academic journals and for public audiences, but the knowledge extracted by the State is rarely disseminated, and when it does become public, it is often through disinformation, indictments, and activist trials.

Social Movement scholars may be asked to directly provide information to the State regarding the movements they study. Our encounters with State repression demonstrate that scholars may possess information that the State seeks, and this may place those scholars in a tenuous position.

Rik Scarce is perhaps the most well-known example of what may happen to scholars studying animal and earth liberation movements. A professor of sociology at Skidmore College and a noted expert on radical environmental movements, Scarse is also the author of the classic book *Eco-Warriors* (2005). He spent five months in jail for contempt of court when he refused to testify to a grand jury in 1990 regarding an investigation of an unsolved break-in at an animal research and testing lab. He refused to testify on the grounds that the confidentiality agreements associated with academic research protected him and his research participants just as shield laws protect journalists.

Years later, Scott DeMuth refused to speak to a grand jury and spent several months in prison because the federal government believed he had information on animal liberation activities. David Pellow was also approached by the FBI in relation to this case. Our refusal to speak to the authorities was a decision we made as activists and as scholars. As activists, we maintain that speaking to investigators about the social movements we are involved in is not a good idea. The government has historically used such investigations to disrupt and neutralize social movements.

As scholars, our decision not to speak to the federal government was an ethical decision because we must honor the confidentiality agreements we have with our research participants. To speak to the authorities would have not only violated the trust that people placed in us but would have also been a breach of our own academic integrity. This would have constituted a blatant disregard of our research participants' privacy and trust. And in our own self-interest, speaking with investigators would have greatly compromised our future research. Even if our research had focused on a social movement we were politically and morally opposed to (for example, a white supremacist movement), we would have taken a similar stance. And even if we had not been conducting this research, we would have taken the same position of non-cooperation as activists.

Although there are examples of and support for scholars who take a firm stance on protecting subject identities, some scholars may willingly give up information they extracted from movements (even though activists in those movements would not willingly have given that information to the State). More subtle, perhaps, are the situations in which scholars studying movements have been asked to serve on government task forces and research bodies. A number of scholars have facilitated and taught trainings sessions with State security forces regarding specific social movements. While this is often done under the pretense of building democratic "pluralism" and mutual understanding, we see such actions as another extension of the science of repression. The ethics of participation in such state-sponsored projects should be weighed heavily by scholars.

Academia as a Free Space for Innovation and Liberation

We have identified the ways in which State repression is applied against social movements and the academy, but we also want to extend the conversation about academic repression to include the ways in which scholars are engaged in repressing social movements and social transformation. Even though scholars engage in social movement research from a position of affinity and solidarity with these movements, good intentions are not enough to mitigate and counter the repressive and exploitive potential of their actions.

Although we have just issued a caution about the roles of the academy and scholars, we also want to highlight the potential that exists in these spaces and positions. There are many real political and social privileges within the academy. Rather than feeling guilty about these privileges, ignoring them, or doing nothing with them, we hope to strategically use these privileges in ways that build communities struggling for justice as well as bridges between these communities and universities. As scholar-activists, we work to connect social science research with grassroots political action. If producing, transmitting, and implementing new knowledge are at the center of what constitutes a social movement, then activist-scholars can play one part in contributing to social change through creating and sharing ideas with our students, colleagues, and the public.

We see our position as scholars and our training in the social sciences as most useful for: 1) providing movements with resources, 2) mutual sharing of knowledge, expertise, skills, and privilege, and 3) collaborating with social movements to produce new knowledge and achieve mutual goals. We see these tasks as part of a broader project of de-centering the university.

As North American university-based scholars, we have considerable resources at our disposal. The university has been a site of social movement origin, coalescence, and mobilization. This has been a space for groups to meet, congregate, and rally. Additionally, scholars have access to resources to bring in speakers on behalf of local movements, to support speaking tours for movements, and also to host conferences. These are a few of the ways in which scholars can use their material privileges to continue to shape universities as "free spaces" and sites of movement innovation. Movements have long shared knowledge with researchers, but scholars have not always been as willing to share the privilege of their position.

This mutual sharing should be extended so that knowledge production and theory creation are not seen as the sole work of academics: The active production by movements should be acknowledged and respected by researchers. Furthermore, this sharing should also include skills and expertise on conducting research, collecting data, and analyzing results. Many movements are already engaged in these practices.

This mutual sharing is the beginning stage of a broader project of continuing collaboration between scholars and activists. The research process has historically been a fraught one-way process with knowledge being transmitted in only one direction. Social movement practitioners and scholars both hold knowledge and theories that can be shared, contested, and tested. Activist research, particularly as it pertains to social movement research, should be a collaborative process.

As activists and scholars studying and working with and within social movements in the U.S. and globally, we have had amazing, mutually beneficial experiences collaborating with groups and movements outside of academia to produce new knowledge with the motivation and goal of social change. These collaborations have included framing issues of power, identifying political opportunities, researching case studies of effective opposition to specific institutions as well as helping to create broader strategies for organizations.

There is a continued need for an engagement with Social Movement Theory; however, this literature can benefit from shifting its gaze to include collaborative research projects that are of substance to movements. This leads us ultimately to more questions than answers. What exactly is activist scholarship? Is it participant observation, collaborative projects, or action research? How is it distinguished from social movement research? Is it research about social movements, or is it the research done by these movements?

This blurs the lines of activist and scholar, participant and observer, and also of social movement scholar and activist-scholar. We have used the last two terms interchangeably, as we do not seek to constrain and contain the boundaries of either. The same should be true for participant and observer in collaborative research. We believe there is more to gain by juxtaposing these categories rather than attempting to create clear definitions and lines between them.

Ultimately, we see activist research as a process, a standpoint, and reflection that incorporate a multitude of epistemologies, methodologies, pedagogies, and praxes. This multitude is bound by a shared motivation and sense of purpose for producing knowledge and practices with the end goal of social change, transformation, and liberation.

We take guidance from Michael Burawoy, a renowned methodologist and public sociologist. Although he uses language that is marked by distinctions between the researcher, who is seen as the sole producer of theory, and participants, who are "observed" and "under study," Burawoy describes a process in which the researcher adopts the conditions and struggles of participants. Burawoy did this himself in the industrial factories of post-colonial Zambia, during the reconstruction of capitalism in Chicago, in the collapse of socialism in Hungary and Soviet Russia, and in the post-Soviet economic situation in Russia (Burawoy, 2009).

This includes a rejection of what Bourdieu calls the "scholastic enclosure" (2000) and signals a commitment to scholarly engagement with the world. Burawoy highlights Theda Skocpol's *States and Social Revolutions* (1979) and Leon Trotsky's *Results and Prospects* (1906/2010) to show that engagement with the world is essential to the scientific process rather than a liability (Burawoy 1998a; Burawoy 1998b). Burawoy calls on researchers to extend their experience to include those of the lives they research, which is to include extended periods of time in their homes, mines, or factories (Burawoy 2009).

Awareness of the destructive means to which one's research might be deployed is an obligation. One of the main questions we must ask is whether we have earned the respect, trust, and consent to study and research with these communities? This should be based on a demonstrated commitment to the cause of social change that the community is working toward. It does not necessarily mean that one needs to always march in lock-step with members of an activist collective but rather that one demonstrates a commitment to working with people to achieve goals even if that entails any number of periodic disagreements and debates. We practice activist-scholarship in this vein, and we would expect any serious activist to do the same.

The most important question that defines the position of activist-scholars is where do our loyalties lie? For us, the answer includes both a critical solidarity with social change movements as well a commitment to solid scholarship and evidence-based practices to be employed in movements for social and economic change and justice.

Burawoy insists that ideals of social and economic justice can be informed by these experiences and that these struggles for social justice can and should drive a scholarly agenda (Burawoy 2009). He cites the example of scholars and the popular struggle in South Africa: "Visiting South Africa in 1990, I was surprised to discover the close connection between sociology and the anti-apartheid struggle. ...While in the United States we were theorizing social movements, in South Africa sociologists were making social movements" (Burawoy 2007: 9). Burawoy concludes *The Extended Case Method* by calling for scholars to undertake "ethnographies of social transformation," in which they seek out, create, and embody alternatives and resistance to injustice (Burawoy 2009: 265).

Bibliography

Bourdieu, Pierre. (2000). *Pascalian meditations*. Cambridge, UK: Polity Press.

Boykoff, Jules. (2007). *Beyond bullets: The suppression of dissent in the United States*. Oakland, CA: AK Press.

Burawoy, Michael. (2009). *The extended case method: Four countries, four decades, four great transformations, and one theoretical tradition*. Berkeley: University of California Press.

Burawoy, Michael. (2007). The field of sociology: Its power and its promise. *Public sociology: Fifteen eminent sociologists debate politics and the profession in the twenty-first century*, pp. 241-258. Berkeley: University of California Press.

Burawoy, Michael. (1998a). Critical sociology: A dialogue between two sciences. *Contemporary Sociology*. *27*(1): 12–20.

Burawoy, Michael. (1998b). The extended case method. *Sociological Theory*. *16*(1): 4–33.

Eyerman, Ron, and Andrew Jamison. (1991). *Social movements: A cognitive approach*. University Park: Pennsylvania State University Press.

Foucault, Michel. (1995). *Discipline and punish: The birth of the prison*. New York: Vintage.

Hale, Charles R. (2001). *What* is *activist research?* Items & issues. Social Science Research Council 2(1-2):13-15.

Kitson, Frank. (2011). *Low intensity operations: Subversion, insurgency and peacekeeping*. London: Faber and Faber.

Lawrence, Ken. (2006). *New state repression*. San Francisco: Tarantula Press.

Marx, Karl, and Friedrich Engels. (1998). *The German ideology*. Oxford: Blackwell.

Morris, Aldon. (2009). Tactical innovation in the civil rights movement. In Goodwin and Jasper (Eds), *Social movement reader*, 2nd ed. Oxford: Blackwell.

Polletta, Francesca. (1999). Free spaces in collective action. *Theory and Society*, *28*: 1–38.

Scarce, Rik. (2005). *Eco-Warriors: Understanding the radical environmental movements*. Walnut Creek, CA: Left Coast Press.

Seidman, Gary. (2001). Guerillas in their midst: Armed struggle in the South African anti-apartheid movement. *Mobilization* 6(2): 111–128.

Skocpol, Theda. (1979). *States and social revolutions*. New York: Cambridge University Press.

Stacey, Judith. (1988). Can there be a feminist ethnography? *Women's Studies International Forum* *11*(1): 21–27.

Trotsky, Leon. (2010). *The permanent revolution & Results and Prospects*. Saddle River, NJ: Red Letter Press.

• CHAPTER TEN •

The College Campus as Panopticon

How Security and Surveillance Are Undermining Free Inquiry

Joe Lewis

In late August of 2001, I moved to the Morningside Heights neighborhood of Manhattan to begin four years of graduate school at Teachers College, Columbia University. This was my first time living in New York City, and I was astounded (and a bit overwhelmed) by all that the city had to offer. One thing in particular struck me about Morningside Heights: the glaring iconography of Columbia University, sometimes called the "Academic Acropolis" of New York City. Columbia's main campus is an imposing array of neo-classical and Renaissance-styled buildings that covers six city blocks. Low Memorial Library, in particular, a National Historic Landmark, has become a kind of emblem for "the elite academy." By any measure, Columbia's main campus is impressive. It is also, essentially, a fortress—situated on a high hill above Harlem and surrounded by heavy wrought-iron fences and gates. As a student and instructor at Teachers College, I liked the idea that our campus lay below the hill and outside the iron fences of the main campus. Metaphorically at least, TC was situated *with* the people of the city (and Harlem in particular), giving its smaller, more humble campus a very different feel and, one hopes, a very different mission—our icons were John Dewey and Maxine Greene, not Herodotus and Cicero.

For me, the location and layout of Columbia's main campus, its stately architecture, its heavy iron fences and gates, even the homeless people squatting around its perimeter, communicated a clear message: an elite class of people worked and resided here; the "riffraff" of the city were not welcome. Ironically, this layout and design bears some striking resemblance to another influential institution in our nation—take away the classic architecture and the manicured lawns and you are left with a minimum security prison (except, of course, that the purpose of a prison is to keep people in, not out).

Less than two weeks after my arrival in New York, the events of September 11th stunned the city. 9/11 had an immediate influence on security everywhere, and Columbia was no exception. Before the attacks, I could access the main campus by passing through an iron gate on 120th Street and climbing two flights of steep stairs. After 9/11, this northern gate, which faced Teachers College and Harlem, was closed and locked with a deadbolt. Now I had to walk the perimeter for several blocks (passing the homeless who sat on Columbia's giant air vents) and enter with an ID card through a gated security checkpoint at 116th Street and Broadway. Back at Teachers College, our own security force became much more vigilant about checking IDs, limiting access to the campus, and monitoring on-campus activity.

Since 9/11, our national understanding of and tolerance for security and surveillance have undergone a seismic shift—the objective now is not only to keep the "riffraff" out, but to scrutinize our own communities and police them for potential "misbehavior." On college and university campuses across the country, a proliferation of surveillance and security systems has pushed us toward a panoptic culture, where students, faculty, and staff are observed and monitored in a variety of ways that tend to promote mistrust, fear, self-surveillance, and, ultimately, self-censorship. Responding to a post–9/11, post–Virginia Tech world, we have allowed a desire for safety and security to bring about a dramatic increase in on-campus and cyber-situated systems of surveillance. Technological advances have led to new and more invasive forms of surveillance, and the U.S. justice system has struggled to keep pace with their legal implications. Meanwhile, the USA PATRIOT Act (Uniting and Strengthening America by Providing Appropriate Tools Required to Intercept and Obstruct Terrorism) has loosened our standards for personal privacy. Particularly in the past ten years, U.S. colleges and universities (like many institutions) have become strikingly more panoptic.

I do not claim that surveillance and security systems have no legitimate uses; nor do I claim that they should be removed completely from college campuses. I argue, instead, that such systems and their uses must themselves be monitored with vigilance. The university is an especially dangerous cultural space to allow panopticism to take hold, since the university exists to nurture—not monitor or censor—free inquiry, open communication, debate, experimentation, and tolerance. Some have argued that this vision of the university is a myth (Althusser, 1971; Bourdieu, 1970/1990; Kaltefleiter and Nagel, 2010). I remain convinced that it can, and must, exist if we are to hold out any hope for a more just social order.

This chapter proceeds in three steps: first, I revisit panopticism and Foucault's analysis of its effects; next, I lay out the various ways in which systems of security and surveillance have increased over the past ten years on college campuses, contributing to a culture of panopticism; finally, I reflect on the damaging influences of a panoptic culture and suggest some ways in which we might keep it in check on college campuses.

Panopticism and Its Postmodern Implications

Jeremy Bentham's classic Panopticon depends on architectural design for its effect. Prison inmates are placed in cells that encircle a central watchtower with reflective windows. This design allows for constant surveillance of each inmate, while simultaneously denying inmates any information whatsoever about their observers ("Who is there?" "How many are there?" "Is anyone there?"). Foucault (1977) argues that Bentham's Panopticon can lead to significant degrees of self-discipline, self-control, and "right" behavior. "Trapped by visibility," the inmate comes to assume that s/he is always being observed and monitored, resulting in "the automatic functioning of a regime of silent discipline" (Macey, 2000, 290). Panopticism, then, is "a form of power that relies not upon overt repression but upon the constant surveillance of a population and 'discipline'. . . of the body" (Macey, 2000, 290). The invisibility and anonymity of the observer serves to decentralize power and make it more effective and efficient. Anyone can serve as *surveiller*, and continual monitoring is not necessary. What's essential is that the prison inmate—or the patient, or the employee, or the student, or the faculty member—*believes* that s/he is being watched.

The availability of video cameras and other forms of electronic surveillance has allowed institutions, both public and private, to produce a panoptic effect in virtually any setting without significantly altering site layout or architectural design. In a postmodern, post–9/11 world, we are *all* called upon to serve as watchers and surveillers. "If you see something, say something," goes the Department of Homeland Security public awareness campaign. Meanwhile, none of us is free from the ubiquitous "gaze." In *Everyday Surveillance*, Staples (2000) outlines four characteristics of postmodern disciplinary practices:

- They are increasingly technology-based, methodical, automatic, and . . . anonymously applied.
- [They] target and treat the body as an object that can be watched, assessed, and manipulated.
- [Their] techniques are often "local," operating in our everyday lives.
- [They] bring wide-ranging populations . . . under scrutiny. (11)

Staples argues that postmodern disciplinary power is "capillary—it expands out, colonizes, and moves to the tiniest reaches of social life" (30). *All* societal spaces are threatened by the capillary movement of disciplinary power, including the supposedly open and freethinking space of a college campus.

Surveillance and Security on College Campuses

Cameras and Other Forms of Electronic Surveillance

Cameras can be placed strategically to allow for monitoring of potential "trouble spots"—parking lots, building entrances, stairways, lobbies, or places where money changes hands. Here, the panoptic effect results in part from a deliberately visible placement of video equipment intended to discourage crime or other forms of "misbehavior." Consider the video camera and monitor that face us as we pay for a cup of coffee at a gas station—we function simultaneously as observed and observer, and our behavior may change, if only slightly. (I tend to stand up straighter when I see myself hunching on a video monitor).

Increasingly smaller cameras have allowed for monitoring that is deliberately *not* visible. The goal here is to catch individuals in the act of "misbehaving," like a highway speed trap. Catching enough perpetrators helps security forces to validate their own existence as crime-stoppers and to deter future perpetrators through the publication of crime records, criminal apprehensions, and security alerts (more on this below). Digital storage systems have allowed for very inexpensive recording, resulting in near-constant visual surveillance and a permanent record.

Like most institutional settings, colleges and universities have dramatically increased camera surveillance over the past ten years. "Many campuses that had 10 to 20 cameras a few years ago, now have 150 to 200," said one security consultant in 2003, four years before the Virginia Tech massacre (Young, 2003, A36). Universities rarely publish detailed information about their security systems, but a look at several university websites suggests that most campuses have dozens, if not hundreds, of strategically placed surveillance cameras (student population and campus size influence this number, of course); in most cases, the total number of on-campus cameras appears to have at least doubled, if not tripled or quadrupled, since 2001.

A perusal of online vendors shows that video surveillance on college campuses is big business. In addition to touting the benefits of crime prevention, online vendors openly advertise the potential to observe student and faculty behavior. One site says, "Surveillance cameras can monitor campus activity, working to ensure that school policies are properly followed" (Video Surveillance, n.d.). A competing site advertises its video system as "the ideal school security system" with "unlimited locations and cameras" and an ability to "view all schools and buildings from a single display" (Video Insight, n.d.).

Many vendors boast of an ability to view off-campus property with cameras that pan and tilt. One surveillance blogger asks: "Is face recognition the next step in school surveillance?" Another writes, "The ideal goal should be to convince [individuals] not to even attempt to do something that is unacceptable" (Camera Security Now, n.d.).

Before moving on to other types of surveillance, I'll emphasize two key points about cameras: (1) the dramatic increase in the total number of on-campus cameras now makes it possible for campus security forces (and administrations) to observe and police most of a college campus; (2) smaller camera size and improved effectiveness have made covert surveillance much more common, which increases the likelihood that surveillance may be employed inappropriately—to monitor employee performance; to observe off-campus behavior; to "profile" students or visitors to campus based on perceived negative attributes (such as race, ethnicity, or style of dress); to scrutinize political protests and demonstrations.

In addition to the now-ubiquitous surveillance camera, many other forms of electronic surveillance are now possible, due to an explosive surge in communication technologies—email, text messaging, social network sites, and cellular phones. All of these communication technologies can be monitored, often without participant knowledge, through various kinds of spyware. Though we tend to associate spyware with the computer-underground world of hackers and crackers, it is employed most commonly by U.S. government officials seeking to monitor suspected "terrorists" and/or political dissidents. Spyware can also be purchased for corporate computer systems (in fact, it is often included as part of a corporate computer package) and can then be used to monitor employee performance. These days, any individual can attach a wallet-sized GPS tracking device to a car or purchase spyware to secretly tap a cell phone.

Not unlike a corporate network of employees, a college community is highly interconnected (MacDonald, 2003). In addition to sharing the physical space of a campus, students, faculty, and staff are linked electronically by a central Internet server. More and more coursework is conducted online, through hybrid and fully online delivery systems. Student and employee records are typically kept in a central database, which is often overseen by a single information-technology department. Electronic ID cards make it possible to monitor an individual's on-campus purchases and the library materials they've checked out. Cell phones that are linked to a campus security alert system can be tracked using GPS technology. As noted above, cell phones can also be tapped. All of this makes it much easier to monitor the activities of individuals or groups in a college or university setting, markedly increasing the degree to which college campuses have become panoptic. Recall that Bentham's Panopticon does not require the presence of an observer to have its effect. The possibility of being watched can result in self-surveillance and self-censorship. To what degree are students and faculty altering their behavior and censoring their political views under the gaze of the panopticon? A panoptic culture undermines free inquiry and open debate even for those who aren't actually being watched. Meanwhile, evidence suggests that college campuses *are* being observed much more routinely than in the past. This is due in part to the improved surveillance technologies described above. It is also due to a substantial loosening of government restrictions on surveillance.

The Policing of Campus Politics

The USA PATRIOT Act has drastically reduced limitations placed on law enforcement officials who may want to monitor on-campus activity. Thinking from the perspective of student civil rights, the *Los Angeles Free Press* points out that the PATRIOT ACT

- broadly expands the official definition of terrorism, so that student groups that engage in certain types of civil disobedience could find themselves labeled as terrorists . . .
- gives law enforcement access to student educational records without probable cause of a crime being committed. . .
- allows law enforcement to conduct secret "sneak and peek" searches of a dorm . . .
- permits the government to monitor Internet traffic and e-mail communications on any Internet service provider without probable cause of a crime being committed . . .
- permits a vast array of information gathering on student political meetings and religious ceremonies . . . (Patriot Act Stands, n.d.)

The PATRIOT Act has led to increases in warrantless searches, wiretappings, detentions without hearing, and ethnic profiling. Government agencies are putting these newly acquired powers to use on college campuses (along with many other locations). In Minnesota, the sheriff of Hennepin County has branded several student groups as possible terrorist threats, including the Anti-Racist Action group, which protests neo-Nazism and white supremacism, and Students Against War, which opposes U.S. militarism (Gruchow, 2003). The work of Shafiqa Ahmadi (2011) demonstrates that Muslim students and scholars have experienced excessive degrees of monitoring and policing from the U.S. government. As a result, the U.S. has lost (and continues to lose) a large pool of talented international students and scholars to more welcoming nations like Canada and Australia.

Ahmadi's work examines key sections of the PATRIOT Act that infringe upon the First and Fourth Amendment rights of Muslim students in U.S. colleges and universities. She argues that intrusive security measures are being used to deny visas and prevent significant numbers of Muslim students and scholars from coming to the U.S. On campus, the effects of surveillance have been palpable and alarming. Formerly active Muslim student groups have disbanded for fear of being branded as terrorist organizations. Contributions to Muslim charitable organizations have declined significantly. There is evidence to suggest that Muslim cultural studies (which is arguably more essential now than ever before) is on the decline due to concerns over increased government surveillance. Ahmadi shows that Muslim students are often afraid to worship and even assemble for fear of scrutiny. She argues that a "false racialization" of Muslims has begun to occur—students who appear to be Middle Eastern or Arab are scrutinized more often and more comprehensively. In the name of safety and security, we have targeted a specific group of people for greater on-campus surveillance. By restricting (or indirectly curtailing) free inquiry, free assembly, and the free exchange of ideas, we deny true academic freedom on our college campuses. The current political atmosphere puts Muslim students and scholars under the microscope (an odious violation of civil rights), but the PATRIOT Act itself does not discriminate. Any of us is susceptible to such scrutiny if our views or actions are deemed suspect at some point in the future.

As silent surveillance and scrutiny increases, a less subtle shift has also occurred in how many campus security and police forces respond to on-campus political protests. Most recently, political protests on the University of California campuses have garnered media attention due to overzealous police response. Linked to the Occupy Wall Street Movement, student groups across California staged on-campus protests and sit-ins to oppose a dramatic hike in state tuition rates. At UC–Berkeley, students attempting to set up a protest site were assailed by campus police wearing riot gear. The police tore down tents, arrested many students, used batons to beat resisting protesters, and fired bean-bag guns into the crowd (Occupy U.C. Berkeley, 2011). At UC-Davis, passive protesters "took a face full of pepper spray at close range from an officer in riot gear" (UC Davis, 2011). These incidents are specific to the campuses, protesters, and police officers involved, but they underscore a general trend in how campus police are responding to student protests—peaceable protests are much more likely to be disbanded; greater numbers of students are being arrested; in many cases, campus (or local) police are responding with greater force and more menacing measures of "crowd control." One explanation for such stepped-up measures of policing is a genuine concern over campus safety following the Virginia Tech shootings of 2007. Like 9/11, this incident has had a chilling effect on how we talk about and practice "safety and security," especially on college campuses.

The Public Performance of Campus Security Alerts

On April 16, 2007, the deadliest massacre by a single gunman in U.S. history occurred on the campus of Virginia Tech University in Blacksburg, Virginia. Thirty-two people were shot dead and 25 were wounded. Within hours of the shootings, a media storm fell upon the campus; we were all permitted to "gaze" at the Virginia Tech community as it struggled through the shock and pain of this horrendous event. Politicians used the incident to appear more "statesmanlike" and/or to push their own political agendas. The incident prompted a review of gun laws, mental healthcare policies, and, of course, campus security systems.

Little has been done to change gun laws since the Virginia Tech incident, but a concern for security and liability has prompted most college administrations to create formal security alert systems on their campuses. Email and text-message alert systems are now commonplace, allowing security forces to keep students and employees informed of campus lockdowns or closures. Some schools have added campus-wide public address systems. As noted above, campus police forces and college administrations are more likely to disband peaceful political protests, and many have ramped up their on-campus police presence and tactics. Security forces have become much more vigilant about widely publicizing "significant" incidents of crime, often through the campus alert system. Below I describe one example from my own institution.

I teach at a small urban university in the Midwest that prides itself on cultivating civic responsibility, social justice, and public service. The primary campus is located in a middle-class neighborhood with a comparatively low crime rate. This neighborhood borders a working-class area of the city with slightly higher crime rates and many more people of color (including recent immigrants). I take the public bus to and from campus, often after dark, and have always felt safe. On campus, we have occasional thefts or cases of vandalism. As in any urban setting, there is violent crime in our city, though it rarely occurs anywhere near our campus. By any reasonable assessment, we are a safe and secure urban learning community with easy access to a wonderfully diverse city (though we are, of course, just as susceptible to a Virginia Tech–like incident as any institution, since such incidents are random acts of lunacy. In my view, they will continue to occur as long as we have easy access to deadly weapons).

Operating under the hyper-vigilant expectations of post–Virginia Tech security, our campus security force publicizes "critical incidents" that have occurred on or near campus, using our new security alert system (its very existence seems to compel its use). My intention here is not to critique the security personnel at my institution—they are professionals who appear to be doing the job they were hired to do. To my knowledge, they have not engaged in any of the excessive police tactics described above. But I do want to offer an analysis of the discourse used in these "critical incident reports," with an eye toward critiquing their damaging effects on our community of learning and the community beyond our campus.

One recent report states: "The suspect is described as a 17–18 year old black male with short hair . . . The group of young men he was with are described as five black males ages 17–18 with short hair." The report reassures us that the police have increased patrols in our neighborhood and that our security staff has widened its patrol to include more off-campus streets. Among the general safety tips offered: "If approached, get away as soon as possible. . . . Ignore and walk away from panhandlers or tell them that you will call the police. . . . Cross the street and go the other direction . . ." This particular incident involved some taunting by a group of young men and one punch.

Though I am not in any way condoning the act described, I question whether it is a "critical incident" that requires campus-wide publicity. I worry that such reports encourage our students (and perhaps our faculty and staff) to fear and steer clear of our neighbors and the people of our city, particularly people of color. The reports tend to focus on race, since this is the physical trait that witnesses most often remember; meanwhile, we know that cross-racial identification is highly unreliable in witness testimony. I can only imagine the emotional tension that such reports must create for students and faculty of color. And I simply do not understand why we need to be shunning the poor—"panhandlers"—particularly at an institution that seeks to cultivate civic responsibility and social justice. Do such security measures serve to undermine the mission and values of our university? I offer it as one example of how campus security alerts may contribute to a culture of fear, suspicion, and mistrust.

The Policing of Faculty

Perhaps the most disconcerting aspect of increased surveillance and security on college campuses is that it directly undermines free inquiry and the free exchange of ideas among scholars. "The preeminence of American Universities may be at risk," argues Jonathan Cole (2003), provost and dean of faculties, emeritus at Columbia University. "[P]olicies that have been implemented in the name of national security . . . threaten to undermine some of the core values of the university . . . . Are we prepared to relinquish personal privacy and academic freedom to secure some vaguely articulated increase in national security?" Cole asks in a talk presented at the University of Chicago (2003).

In *Academic Repression: Reflections from the Academic Industrial Complex* (Nocella, Best, and McLaren,2010), an array of critical theorists and scholars delineate how they and others have become the target of academic repression through government surveillance; excessive media scrutiny; attempts to censor faculty free speech; revocation of visas; denial or revocation of tenure; revocation of invited speaking engagements; and in some cases, arrest and imprisonment. Among the scholars discussed or represented: Sami Al-Arian, Bill Ayers, Steven Best, Sarahjane Blum, Noam Chomsky, Ward Churchill, Norm Finkelstein, Victoria Fontan, Henry Giroux, Rashid Khalidi, Steve Kurtz, Edward Said, and Rik Scarce. Each case has its own complications and nuances. Taken collectively, they are a chilling indictment of the current state of free inquiry in our university system. The editors of *Academic Repression*—Anthony Nocella II, Steven Best, and Peter McLaren—question whether academic freedom can exist in a political atmosphere that countenances and even seems to encourage covert surveillance, summary judgment, and the bypassing of established university processes for peer review and shared governance.

Meanwhile, we simply do not know how deeply this panoptic culture has influenced faculty who are not (at least currently) the target of close scrutiny. "To what extent are intellectuals who do not consider themselves particularly radical or progressive now policing themselves?" ask Kaltefleiter and Nagel (2010). "To what degree have they internalized social censorship . . . so that they no longer enjoy or even give thought to the erosion of academic 'freedom'?" (407). I'll admit (as a non-tenured, assistant professor) that I engage in various degrees of self-censorship. In writing this essay, I've asked myself: Am I being too critical of my own institution? Am I burning bridges with the administration? Should I be writing about this topic at all, especially at this point in my career?

Cary Nelson (2010), president of the American Association of University Professors, delineates a series of current forces that he believes threaten our higher education system:

1. The managerial model that now dominates the corporate university.
2. The massive reliance on contingent faculty, a trend that has doubled over the last thirty years . . .
3. The loss of faculty vigilance over and understanding of the relationship between shared governance and academic freedom . . .
4. The renewed culture wars waged by the Right to deprive faculty of academic freedom . . . (468–469).

From Nelson's perspective, free inquiry has been eroded not only because administrations have become more vigilant about scrutinizing their faculties, but also because faculties have become less vigilant about protecting their own academic freedom and systems of shared governance. Richard Kahn (2010) holds out hope for the academy as a space for inclusion, but also underscores the need for continual critique:

> Higher education *does* allow for a level of debate and ideological inclusion that is certainly welcome and, depending on where one lives, may be a veritable sanctuary for faculty, students, and citizens who are interested in pushing beyond expected social norms in their research agendas and lifestyles . . . . I am happy to identify as an academic . . . but this does not mean that I act as an apologist . . . [T]he privileges that come with such status . . . place an ethical demand upon me as a scholar—that I undertake an ongoing critique of the limitations and contradictions of my place in society as a professional academic. (201)

Like Nelson, Kahn argues that academics have a responsibility to remain vigilant about protecting academic freedom and a duty to protest when it is being threatened.

The Dangers of the Panoptic Culture

Of course we want our college campuses to be physically safe spaces. But a functioning community of learning also requires emotional safety and trust. Meaningful education can only occur when students are free to take risks, make mistakes, try out new ideas, experiment with different forms of expression, learn to thoughtfully critique the status quo, and possibly conceive of better, more just, social contracts. A community that constantly surveils its members, that asks and expects them to surveil one another, undermines such goals. This surveillance begins from an assumption that "unacceptable behavior" *is* occurring and must be stopped. Surveillance, by its very nature, makes all of us suspect.

We also want a student population that is actively engaged with the broader community—through service learning, volunteer opportunities, internships, and political movements. I am proud of the various ways in which my own institution encourages and expects community engagement. Such engagement is undermined when students are prompted to fear and mistrust their neighbors and fellow citizens. We need campus policies that keep students reasonably safe, while still fostering meaningful involvement in the broader community.

Extensive surveillance and excessive security measures also quell any real challenge to the existing power structure. From an educational perspective, this is dangerous. A college campus is meant to be a space that challenges social and political traditions and tenets. Protests, petitions, sit-ins, and demonstrations may cause headaches for college administrations and campus security forces, but they are a critical facet of a college education in a democracy, and an excellent opportunity for experiential learning.

During my four years as an undergraduate at a liberal-arts college in the middle of Iowa, I participated in many political protests, campaigns, and sit-ins, but I especially remember a "Divest from South Africa" sit-in that occurred outside the college president's office. It stands out in my memory because the president came out on the lawn, sat down, and chatted with us about the issue for over an hour. We learned that he was well-informed about South African politics, agreed with our goals, but disagreed with our strategies. We learned that he respected the political protest as a meaningful expression of opinion that deserved his thoughtful attention. Speaking directly with us was a brilliant political move—the sit-in ended soon afterward, since all of us felt that we'd been heard. As a continued symbol of protest, students constructed a small shanty-town on the campus quad and took turns squatting in it for several weeks. All of this occurred without threatening anyone's safety or disrupting anyone's education. Imagine if, instead, the local police had been summoned to arrest protesting students and dismantle the illegal shanty-town. I'll be the first to admit that the political impact of these activities was extremely limited, and there were (as there almost always are) some hypocritical elements to our protests. But they were important forays into the political process that led many of us to more meaningful and sustained participation after graduation.

In our post–9/11 hysteria over "terrorism," we seem to have forgotten that political protests are an essential part of our country's democratic history—this is how women got the right to vote; how people of color achieved civil rights (at least legally); how U.S. participation in the Vietnam War came to an end. Protest is also an essential part of our future—the GLBT community will gain its civil rights through political protest; public workers in Wisconsin and Ohio will regain their right to collective bargaining through political protest; the Occupy Movement may finally influence our nation's unjust social contract through political protest. Such movements *belong* on college campuses, where students and scholars should actively participate in open debate and the free exchange of ideas.

Most troubling is the potentially destructive effects that extensive surveillance may have on free inquiry among the community of scholars. "Whenever a controversial scholar or an emerging area of study is attacked," writes Michael Berube (2010), "the chilling message goes out to thousands of instructors, adjuncts, and graduate students: *Watch what you say. Don't offend your students—or their parents. And above all, be careful about the kind of company you keep*" (4). Giroux (2009) argues that such attacks are explicitly intended to "undermine the university as a public sphere that educates students as critically engaged and responsible citizens in the larger global context" (130).

Keeping the Panoptic Culture in Check

In an effort to limit the dangers of the panoptic culture, while still allowing for sensible measures of safety and security on college campuses, I offer the following concrete suggestions:

- Colleges and universities should have explicit surveillance policies, written and reviewed by faculty, students, and administrators through a process of shared governance. Such policies should lay out "reasonable" security measures (as defined by the particular community) and prohibit forms of surveillance deemed inappropriate or unethical. If I were to participate in writing such a policy, I would require that community members be given complete information about the types of surveillance used on campus and the location of all surveillance cameras.
- Security alerts and critical incident reports should be used sparingly. While it may be appropriate to inform students about the location and nature of serious crimes, reports should seek to accurately inform the community without inciting mistrust or paranoia. Descriptions of race and class should be avoided, since they have limited impact on safety and contribute to a culture of suspicion.
- Campus security personnel should be trained in how to appropriately respond to on-campus political protests. The goal should be to keep the community physically safe without undermining free speech and open debate. Extreme measures of crowd control should be prohibited, except where violent protest is occurring. To the degree possible, administrators, faculty, and students should work together to define "acceptable" and "unacceptable" forms of on-campus protest and response to protest (with the recognition that civil disobedience is an important form of protest, which may make full agreement impractical and undesirable).
- College and university administrations should directly challenge the constitutionality of the PATRIOT Act by refusing to allow government agencies to inappropriately and indiscriminately surveil their students and faculty.

Bibliography

Ahmadi, S. (2011). The erosion of civil rights: Exploring the effects of the Patriot Act on Muslims in American higher education. *Rutgers Race and Law Review, 12*(1).

Althusser, L. (1971). Ideology and ideological state apparatuses: Notes towards an investigation. In Ben Brewster (Trans.), *Lenin and philosophy and other essays*. New York: Monthly Review Press.

Berube, M. (2010). The company we keep. Foreword to Anthony J. Nocella II, Steven Best, and Peter McLaren (Eds.), *Academic repression: Reflections from the academic industrial complex*, pp. 1-4. Oakland, CA: AK Press.

Bourdieu, P. (1970/1990). *Reproduction in Society, Education, and Culture* (Richard Nice, trans.) London: Sage.

Camera Security Now. (n.d.). http://www.camerasecuritynow.com

Cole, J. R. (2003, Summer). The Patriot Act on campus: Defending the university post-9/11. *Boston Review: A Political and Literary Forum*. http://bostonreview.net/BR28.3/cole.html

Foucault, M. (1977). *Discipline and punish: The birth of the prison* (Alan Sheridan, trans.). New York: Vintage Books.

Giroux, H. (2009). *Youth in a suspect society: Democracy or disposability?* New York: Palgrave Macmillan.

Gruchow, M. (2003). Patriot Act divides our state's citizens. *The Collegian* Online Edition. http://www.sdsucollegian.com/2003/10/23/patriot-act-divides-our-states-citizens/

Kahn, R. (2010). Operation Get Fired: A chronicle of the academic repression of radical environmentalist and animal rights advocate-scholars. In Anthony J. Nocella II, Steven Best, and Peter McLaren (Eds.), *Academic repression: Reflections from the academic industrial complex*, pp. 200–215. Oakland, CA: AK Press.

Kaltefleiter, C. K., and Nagel, M. E. (2010). The carceral society: From the prison tower to the ivory tower." In Anthony J. Nocella II, Steven Best, and Peter McLaren (Eds.), *Academic repression: Reflections from the academic industrial complex*, pp. 400–414. Oakland. CA: AK Press.

Macey, D. (2000). *The Penguin dictionary of critical theory*. London: Penguin Books.

MacDonald, M. (2003, November 24). Patriot Act stifles dissent on campus. *Baltimore Sun*. http://articles.baltimoresun.com/2003-11-24/news/0311240117_1_student-groups-student-information-college-campuses

Nelson, C. (2010).The three-legged stool: Shared governance, academic freedom, and tenure. In Anthony J. Nocella II, Steven Best, and Peter McLaren (Eds.), *Academic repression: Reflections from the academic industrial complex*, pp. 468–478. Oakland, CA: AK Press.

Nocella II, A. J., Best, S., and McLaren, P. (Eds.). (2010). *AcademicrRepression: Reflections from the academic industrial complex*. Oakland, CA: AK Press.

Occupy U.C. Berkeley protesters face violent confrontation with campus police (with video). (2011, November 10). *Huffington Post Online*. http://www.huffingtonpost.com/2011/11/10/occupy-uc-berkeley-police_n_1086195.html

Patriot Act stands over students (n.d.). *Los Angeles Free Press*. http://www.lafpas.info/?p=7127

Staples, W. G. (2000). *Everyday surveillance: Vigilance and visibility in postmodern life*. Lanham, MD: Rowman & Littlefield.

UC Davis pepper spray video at Occupy protest launches probe by university. (2011, November 19). *Huffington Post Online*.http://www.huffingtonpost.com/2011/11/19/uc-davis-pepper-spray-video_n_1103075.html

Video Insight: Video Digital Management Systems (n.d.). http://www.video-insight.com/Education/Education-Overview.aspx

Video Surveillance. (n.d.) http://www.videosurveillance.com/campus.asp

Young, J. R. (2003, June 13). Smile! You're on campus camera. *Chronicle of Higher Education*, p. A36.

PART III

From Defending Public Education to the Occupy Movement

CHAPTER ELEVEN

Militant Privatization

The UC–Davis Pepper-Spray Incident

Sarah Augusto and Julie Setele

On November 18, 2011, police in full riot gear were called in by Chancellor Linda Katehi to disperse a peaceful student protest at the University of California–Davis. In what has now become international news, one officer used weapons-grade pepper spray at close range on seated students. In this chapter, we will examine the UC–Davis pepper-spray incident in the context of the privatization and marketization of the university and the militarization of the police. As public funding for higher education continues to decline, the University of California has increasingly turned to private funding sources, and in doing so, it is becoming less like a public institution and more like a private corporation. UC–Davis professor Laura Grindstaff argues that "privatization is the substitution of private for public sources of support, whereas marketization is the repurposing of the university to walk, talk, and act like a profit-seeking commodity. The former impacts who has access to the university (who gets in?), whereas the latter shapes what sort of university we become (what are we getting in *to*?)" (Grindstaff, 2012). Grindstaff is one of many who question this trajectory, and since the fall of 2009 students, faculty, staff, and other UC community members have organized against the university's shift from public to private. These mobilizations are often met with police repression, and there have been numerous instances of police violence against protesters on UC campuses. This chapter will examine the connections between the privatization of the university, the policing of protests against this trend, and the broader process of the militarization of the police.

The UC–Davis Pepper-Spray Incident

On November 17th, the day before the pepper-spray incident, students and other community members established an occupation of the campus quad as part of an ongoing series of protests against the privatization of the university and the resulting budget cuts and fee increases. The UC–Davis occupation was also a show of solidarity with the Occupy Wall Street movement and a response to the police violence that occurred at UC–Berkeley on November 9th, when officers beat students and faculty with batons as the protesters linked arms to defend the tents they had erected for their own campus occupation. One UC–Berkeley graduate student sustained a broken rib and many others were injured or arrested. Berkeley Chancellor Robert Birgeneau later issued a statement claiming that "linking arms and forming a human chain to prevent the police from gaining access to the tents . . . is not non-violent civil disobedience" (Birgeneau, 2011).

On the morning of November 18th, students occupying the UC–Davis quad received a message from Chancellor Katehi informing them that the encampment violated health and safety regulations and that the tents would be removed if students failed to take them down by 3:00 p.m. that day. Some students chose to take down their tents to avoid confrontation, but some refused. Several tents were still up when police arrived later that afternoon and a small crowd of students watched as a line of riot police moved slowly across the quad toward what remained of the encampment. The police line stopped a few hundred yards from the remaining tents and issued a dispersal order, giving students three minutes to leave the area. Those who chose to remain gathered in a circle around the tents and officers began to pull people out of line and to arrest them, leaving them cuffed and sitting on the ground.

A few students sat down on the stone walkway that cuts across the quad. Though they could easily have walked around them, the police asked the sitting students to move and they refused. Several other students joined them and linked arms to form a line across the walkway while a large crowd of onlookers gathered around them, chanting "you use weapons, we use our voice!" When the seated students continued to refuse to move, one of the officers walked back and forth along the line, coating the students with a thick layer of orange pepper spray. Some students were sprayed directly in the face. The police began to attempt to pull them apart and arrested those they could separate, shoving the protesters onto the ground as they tried to wipe the pepper spray from their faces. By this time a large crowd had surrounded the police, who retreated with the ten students whom they had managed to arrest. The crowd followed the police to the edge of the quad, chanting "you can go!"

The university provided no medical assistance for the students who had been pepper sprayed. Instead, fellow activists trained in first aid helped to wash their eyes and provided advice about how to ease discomfort and prevent transferring the chemicals to the home environment. Two people who were pepper sprayed had to be taken to the emergency room and one student who was arrested sustained nerve damage as a result of having been cuffed too tightly.

The events of November 18th were widely publicized and many in the UC community and beyond were outraged at the extreme force used against peaceful protesters. The next afternoon, Chancellor Katehi held a press conference in a small room on campus that was scheduled to last an hour. Demanding their voices be heard, students gathered outside during the event. By the time it was over, a crowd of hundreds had amassed. Katehi initially refused to leave the building, allegedly attempting to give the media the impression that she was trapped inside. After several hours, the protesters agreed to form a gap and sat in silence as the chancellor exited the building through the crowd. Now commonly referred to as the "walk of shame," the event was captured on video and circulated widely in the media. The following Monday, a rally on the UC–Davis campus drew thousands of supporters; afterwards students re-established the occupation on the quad, erecting dozens of tents and a giant geodesic dome structure. The following week students occupied Dutton Hall, the administrative building that houses student accounts and financial aid. For the remainder of the quarter, students, faculty, and other community members maintained the occupation of the quad and Dutton Hall. At these reclaimed spaces, occupiers hosted general assemblies, teach-ins, film screenings, and other events.

Race, Policing, and the Pepper-Spray Incident

Many have suggested that the reason the pepper-spray incident at UC–Davis drew so much media attention and provoked such public outrage was because it involved mostly privileged, white, middle-class students. This type of police violence is by no means extraordinary or new—it happens in poor communities and communities of color with regularity. Daniel Filler, a criminal law expert at Drexel University, argues that the militarization of American police has increased drastically since September 11th, 2001. However, because marginalized groups are disproportionately affected by these changes, they have gone largely unnoticed by the general public until the pepper-spray incident:

> As police culture has shifted, the police seem to have forgotten—or perhaps never realized—that these aggressive policies are really only politically sustainable when they're employed against disempowered communities. Airport profiling is OK for Muslims. Random street searches are OK for African-Americans. And seat-of-the-pants immigration detentions are fine for Mexicans. But you can't engage in invasive stops, searches and unsubstantiated de-

> tentions of, and attacks on, white folks without other white people getting agitated. What's good for the goose has never been good for the gander. That's why the UC Davis video is so powerful. (Filler, 2011)

UC–Davis graduate student Breeze Harper (2011) makes a similar argument in a video post on her blog:

> Why is it that it takes police violence against seemingly "peaceful" and "non-threatening" white students at a predominantly white university campus for the mainstream to suddenly "wake up" to the police and state sanctioned violence that us brown, black, red, and/or "Muslim" folk have been trying to get mainstream America to give a shit about for decades? . . . Can we please have a conversation about how white bodies/white campuses/white middle class spaces are almost always constructed more as "innocent" and "non-threatening" than non-white racialized people who are collectively seen as "threatening," even if they are peaceful?

In response to these arguments, many UC–Davis activists have pointed out that the students who were pepper sprayed were not exclusively white. However, the UC–Davis campus is certainly a privileged space, and all students regardless of race are therefore likely to be read as privileged by media producers and consumers. UC–Riverside professor Dylan Rodriguez (2011) makes this point in an article posted on Reclaim UC, a blog that covers the crisis in public education and the UC protests:

> We must be willing to acknowledge that the political and moral indignation over such excessive displays of nonlethal state violence (that is, pepper spray rather than bullets) is quite openly amplified when such tactics are used on white students, in the space of historically white institutions . . . . There is something structurally white supremacist about this very indignation: the political expressions of outrage and institutional shaming over the spectacle at UC Davis are fueled by an over-identification with (historically white) university campuses as places of presumed innocence, wherein enrolled and employed (white) bodies are also presumed to presume innocence. If we are to be honest in this moment, we must also recognize that many have only now been provoked into talking and thinking critically about the violence of the police because they are moved to defend the presumption of white bodily and spatial innocence, and are not necessarily concerned about police violence against those whose bodies (and the spaces they inhabit) are presumed guilty, or something close to it.

Regardless of whether the media attention and public outcry were linked to the race and class of the students involved, research shows that police violence and discrimination is disproportionately perpetrated against people of color and that these abuses are rarely reported in mainstream media. For example, a 2007 report from the United Nations Committee on the Elimination of Racial Discrimination shows that excessive force is disproportionately exercised against people of color in the United States. Weitzer and Tuch (2004) found that blacks and Hispanics reported more experiences with police misconduct than whites and were more likely to encounter repeated experiences with police abuse.

The pepper spray incident is therefore also disturbing because it highlights the discrepancy between the media fanfare around police violence against college students in a privileged space and the relative radio silence on systemic police brutality, discrimination and harassment, especially in poor communities of color. The former is viewed as shocking and worthy of a spot on the nightly news. The latter is not surprising and does not make a good news story because it happens all the time. It is carried out in countless instances in diverse locations rather than one spectacular moment, such as the pepper spraying of UC Davis students by campus police officer LT. John Pike, caught on video to be broadcast around the world. As the following section will demonstrate, the issues of race and systemic problems with policing have been largely absent from the administration's response to the pepper-spray incident.

Seeking Accountability

Immediately after the pepper-spray incident, while videos of the events spread like wildfire through social media and mainstream news outlets, UC–Davis administrators began issuing statements clarifying the university's official position on what had occurred and why. These statements shifted in tone as the public outcry grew. In an email to the UC–Davis community on the night of the incident, Chancellor Katehi justified the removal of the protesters' encampment on health and safety grounds, arguing that she had "no option but to ask the police to assist in [the protesters'] removal." Her statement continued:

> While we have appreciated the peaceful and respectful tone of the demonstrations during the week, the encampment raised serious health and safety concerns, and the resources required to supervise this encampment could not be sustained, especially in these very tight economic times when our resources must support our core academic mission. (Katehi, 2011)

How, exactly, deploying weapons-grade chemical agents on seated protesters promotes the health and safety of the campus and its members remains unclear, as do the economic benefits of responding to protesters with such a demonstration of police force.

The Davis Enterprise has reported that the UC spent about $100,000 on "crisis communications consulting" following the incident. These costs, along with the sizeable sum being paid to the consulting firm Kroll, which is spearheading the UC-wide "fact-finding" investigation into the incident, are covered by the UC's risk insurance policy (Golden, 2012a). Nevertheless, the irony of the university paying consultants over $300 per hour to investigate its own wrongdoings at a time when student fees have more than tripled over the past decade is not lost on the protesters (UC Davis Facts, 2011; CBS Sacramento, 2012).

Amid repeated questions about who gave the order to use pepper spray, UC–Davis Police Chief Annette Spicuzza informed the press that the decision was made by an officer at the scene, not by her or Katehi (KCRA, 2011). Spicuzza justified the police actions, claiming that the protesters had surrounded the officers and were preventing them from leaving: "The students had encircled the officers. . . . They needed to exit. They were looking to leave but were unable to get out" (Pringle, Quinones, and Grad, 2011). As video footage shows, however, Officer John Pike exited the circle by stepping over the seated students before proceeding to douse their faces with pepper spray. Pending investigation, Spicuzza was placed on paid administrative leave, along with two officers, including John Pike (Golden, 2011a).

From the chancellor's first message distributed to campus members on November 18th, Katehi has called for "dialogue" about what happened. Toward that end, her office has organized numerous "town hall" meetings for faculty, staff, and students to listen to statements by Katehi and others in the administration, and to ask questions. Video of one of these meetings was streamed live online, allowing geographically distant members of the campus community, including one of us (Setele), to participate. However, the structure of the meetings was determined by university administration, leaving one professor to reflect, "it was not a format for dialogue; it was a format for propaganda."

A "fact sheet" distributed to attendees at a town hall meeting on November 29th and emailed to faculty on December 1st (UC Davis Chancellor's Office, 2011) contained so many controversial statements that several faculty members independently organized to produce both an annotated version of the administration's document and an original document countering the administration's version of the "facts" (Antonelli et al., 2011b; Clover, Ferguson, Wittman, and Cox, 2011). The counter-fact sheet questioned the veracity of the administration's "fact sheet," saying the original "document's omissions and representations offer implications of responsibility that are neither proper nor accurate" (Clover et al., 2011). In the official fact sheet and elsewhere, Katehi has claimed "full responsibility" for the pepper-spray incident, while simultaneously arguing that she does not directly oversee the campus police department and has no authority to order campus police to act (Sakash, 2011). In a separate blog post, one of the authors, UC–Davis professor David M. Wittman, pointed out the logical flaws in this argument:

> There is a description [in the administration's "fact sheet"] of the chain of command of the UCD police which goes up to the Vice Chancellor level but then, curiously, stops without mentioning that the Vice Chancellor is overseen by the Chancellor. It is as if an American president took "full responsibility" for a military disaster but then issued a "fact sheet" stating that the military is overseen by the Joint Chiefs of Staff, who in turn are overseen by Defense Secretary Leon Panetta. It's a laughably transparent attempt to avoid actually taking responsibility. (Wittman, 2012)

For Professor Wittman, the administration's "fact sheet" was the last straw that compelled him to take action: he writes, "the selection of facts was so incredibly biased that one could come to no other conclusion than that our own administration was trying to bullshit us" (Wittman, 2012).

Both counter documents criticize the discursive environment established by the administration: ". . . beyond factual inadequacies and their consequences for an understanding of the events themselves, the 'fact sheet' (and its updated companion) testifies to a deliberate strategy of rhetorical management which we do not believe is consonant with leadership of this university" (Clover et al., 2011). A poster on Bicycle Barricade, an anti-privatization blog, made similar observations:

> The rhetoric employed by the Katehi regime is designed to delay, displace, and eventually dissipate the administration's accountability for the violence of November 18. Since that date, UC Davis administration has deployed a therapeutic discourse which seeks to shift the focus from accountability (negatively construed as "blame") to healing and moving forward. (Bicycle Barricade, 2011b)

A therapeutic discourse is on display in the latest issue of *UC Davis Magazine*, which was mailed to UC Davis alumni, faculty, and students' parents over the winter break. Though the cover does not mention the pepper-spray incident, the first story in the magazine, "Campus Protests: Continuing the Dialogue," emphasizes the administration's focus on moving forward. The article begins, "It is a time for listening and healing at UC Davis. And the conversation has begun" (Campus Protests, 2012, 2). While few would argue with the goal of increased dialogue, many at UC–Davis have found that the administration repeatedly espouses positive rhetoric about promoting dialogue and building community, but its actions are more limited, often falling short of producing substantive change (Setele and Grindstaff, 2011).

The case of Jerika Heinze, for example, reflects poorly on the university administration's commitment to genuine dialogue. Heinze is a UC–Davis undergraduate who suffered respiratory problems following exposure to pepper spray on November 18th. According to UC–Davis professor Bob Ostertag (2011b), Heinze repeatedly contacted Chancellor Katehi by email and phone between late November and early December, asking to speak with her about the pepper-spray incident, but received no response. Finally, on December 14th, Heinze tracked down Katehi in a hallway of the Capitol building in Sacramento where the Chancellor was testifying about the pepper-spray incident.

> "I've been contacting you every single day, calling your office, sending you emails, and your assistants said they passed the messages along," Jerika told the Chancellor. "You've never responded to me."
>
> "Well, I've asked them to set up an appointment with you," Katehi replied.
>
> "No, you haven't," Heinze insisted. "You absolutely have not." (Ostertag 2011b)

As news of this interaction spread, Katehi's office contacted Heinze, setting up an appointment for her to meet with the chancellor and insisting that a thorough investigation would be conducted to determine the details of this communication failure. Ostertag (2011b) writes,

> Thus to the five investigations of the pepper spray incident already underway, a sixth investigation was launched to determine what happens to communications from students once they arrive in the Chancellor's office. It is my understanding that five people work in this office. Hopefully it will not require too much tax payer money to get to the bottom of this.

When they finally sat down for a meeting, Katehi "listened respectfully" as Heinze discussed her struggles to attain higher education without any family support. When Heinze began to ask Katehi about the pepper-spray incident, however, the chancellor ended the meeting, saying, "This has become an interrogation" (Ostertag, 2011b).

While the university attempted to spin its own version of the pepper-spray incident, other UC–Davis employees, including faculty, students, and staff, joined the discussion, publishing to social media sites their own descriptions and analyses of the events, along with recommendations for moving forward. Where the official statements from UC–Davis and UC administrators tended to justify the UC–Davis police actions and deflect responsibility for pepper spraying seated protesters, the unofficial critiques overwhelmingly called on Katehi to take responsibility and answer key questions about what the police had been ordered to do and by whom. Among those blazing the trail, Nathan Brown, an assistant professor of English at UC–Davis, posted an open letter holding Chancellor Katehi responsible for the police actions and demanding her "immediate resignation" (2011). Though Katehi has claimed that she ordered police to remove tents, but not to use force (UC Davis Dateline, 2011b; Stanton, 2011), Brown argued that the police actions were foreseeable, especially given the police violence against protesters at UC–Berkeley the previous week.

> You are responsible for it because this is what happens when UC Chancellors order police onto our campuses to disperse peaceful protesters: students get hurt. Faculty get hurt . . . . The fact is: the administration of UC campuses systematically uses police brutality to terrorize students and faculty, to crush political dissent on our campuses, and to suppress free speech and peaceful assembly. (Brown, 2011)

This is not the first time the UC administration has been accused of attempting to suppress student activism. In the spring of 2011, a request filed with UC–Davis under the California Public Records Act revealed the existence of the "Student Activist Response Team." Composed of administrators and staff from various campus offices, the team had been monitoring student activists' online activities, including Facebook pages and blogs, as well as sending team members to monitor meetings and actions on campus. The administration insists that the purpose of the team is not to spy on campus activists but rather to ensure that they are able to safely carry out their right to protest. Student activists have countered this assertion by pointing out that if the purpose of the team were really to support student protest, they would have at least informed students that the team existed.

Though the university administration is powerful, a number of factors combine to thwart administrators' attempts to control the narrative about what occurs on campus. The ubiquity of cell phone cameras, blogs, and other tools of self-publication and distribution contribute to a system of "sousveillance," in which average citizens can monitor others, including those in positions of power, and share that information widely (Ganascia, 2010). From the Rodney King beating to the UC–Davis pepper-spray incident, video records of wrongdoing by police officers provide legitimacy and power, albeit limited, to the victims of police violence. In a university setting, individuals have the additional power and privilege of being formally trained in critical thinking skills. Professor Nathan Brown identified this in his open letter to the chancellor:

> You should bear in mind that when you send emails to the UC Davis community, you address a body of faculty and students who are well trained to see through rhetoric that evinces care for students while implicitly threatening them. I see through your rhetoric very clearly. You also write to a campus community that knows how to speak truth to power. That is what I am doing. (Brown, 2011)

Despite his untenured, and thus insecure, faculty position, Brown joined in solidarity with the individuals who were pepper sprayed on November 18th and everyone else who faces police violence in the future for engaging in political actions on campus. Members of the campus community who speak out use their various privileges to question the normality of police violence. The increasing privatization and marketization of universities, however, threatens individuals' willingness to express their criticisms of university administrations.

Katehi has declined to step down as chancellor despite repeated demands from many students, faculty, and alumni. The executive board of the Davis Faculty Association demanded her resignation, stating, "Given the recent use of excessive force by police against 'occupy' protestors at UC Berkeley and elsewhere, the Chancellor must have anticipated that, by authorizing police action, she was effectively authorizing their use of excessive force against peaceful UCD student protestors" (Davis Faculty Association, 2011a). The majority of faculty in several academic departments and programs, including Physics, English, Comparative Literature, and Middle East/South Asia Studies, joined their voices to those demanding Katehi's resignation (Davis Faculty Association, 2011b). Many more academic departments and programs, including Sociology and Women and Gender Studies, condemned the use of pepper spray and demanded a thorough investigation of the incident (Mama et al., 2011; McCarthy et al., 2011). Some of the groups advocating Katehi's resignation, including the Davis Faculty Association Board, English Department, Comparative Literature Department, and Occupy UC Davis General Assembly, also demanded the disbanding of the university police department.

At the rally held on the quad on November 21st, hundreds of people carried signs that read "Katehi resign" or "I have 99 problems and Katehi is 1." Many speakers demanded Katehi's resignation, receiving loud cheers and applause. When Katehi addressed the rally, after waiting in line like anyone who wanted to speak, she expressed her apologies for what happened the previous week. Katehi said, "If you think you don't want to be students in a university like we had on Friday, I'm just telling you, I don't want to be the chancellor of the university we had on Friday" (UC Davis Dateline, 2011a; AggieTV, 2011). Many in the crowd applauded, some yelled out "resign!" and others shushed the yellers.

Within ten days of the pepper-spray incident, over 100,000 people signed an online petition calling for Katehi's resignation (Buscho, 2011). Not everyone agreed: Dateline, the UC Davis news service, mentioned Buscho's petition without linking to it, instead linking to a counter-petition, which received little over 100 signatures in support of the chancellor (UC Davis Dateline, 2011c; Carolina, 2011). Many faculty members, especially those in the physical sciences, expressed their opposition to calls for Katehi's resignation (see, for example: Adejunmobi et al., 2011; Golden, 2011c; Leal et al., 2011; Davis Faculty Association, 2011c).

The UC–Davis Academic Senate voted in mid-February 2012 on several motions expressing its members' confidence, or lack thereof, in Katehi's leadership. With only 37 percent of eligible faculty voting, the Senate rejected a motion of no-confidence, and passed motions expressing confidence in Katehi and disapproval of the use of police force against non-violent protesters (Branan, 2012). The implications of these motions are unclear, however, as their language has been criticized for its ambiguity (Meranze, 2012; Bicycle Barricade, 2012).

As this goes to print, five investigations are underway to examine what happened on November 18th and why. Two of these investigations have been criticized for not being as "independent" as they claim to be (Price, 2011; Bicycle Barricade, 2011a). UC–Davis Vice Chancellor John Meyer leads one investigation despite his role overseeing the campus police department. Another investigation, conducted out of the UC's Office of the President, is led by Kroll, a security consulting firm run by Bill Bratton, former chief of police in Los Angeles, New York, and Boston. In a letter opposing Bratton's nomination, California Scholars for Academic Freedom note Bratton's history as a police chief who oversaw incidents of police brutality:

> [Bratton] specializes in training police to deploy space-cordoning and crowd-control methods that are less violent, but which end up eliminating the idea of public space and public visibility altogether. This kind of vision is problematic for public spaces, but utterly incompatible with the notion of common spaces in public universities. More importantly, such an investigation should be led by those whose interests were directly affected by such police violence, and those whom this University is designed to serve, namely its students and faculty. (California Scholars for Academic Freedom, 2011)

While Bratton's particular policing ideology worries many (Golden, 2011b; Gaynor, 2011; King, 2011), there are also conflicts of interest between Bratton's organization, Kroll, and the UC. The Council of UC Faculty Associations also protested Bratton's appointment:

> We already know that Kroll has provided security services to at least three UC campuses for the past several years. This in itself would disqualify Mr. Bratton from participating in the investigation you propose, even if the role of Kroll and its affiliated companies in defending the financial sector against [Occupy Wall Street] did not raise further questions about its pro-Wall Street and pro-privatization bias. (Meister, 2011)

The work of Kroll's parent company—Altegrity—and its parent company—Providence Equity Partners—indicates additional conflicts of interest beyond Kroll's connections to the UC. UC President Mark Yudof has dismissed these concerns as unfounded (UCLA Faculty Association, 2011), but the choice of Kroll strikes many as insensitive. As one journalist wrote, "those students who were pepper sprayed in the face were protesting, in large part, the encroaching privatization of the university. Picking a company with Kroll's corporate entanglements to conduct the investigation is incredibly tactless at best" (Price, 2011).

Katehi called November 18th "an aberration" and describes herself as "shocked" by the incident (Katehi, 2012), but this framing overlooks the structural and systemic roots of what occurred. UC–Riverside professor Dylan Rodriguez (2011) argues that what happened at UC–Davis was not an "aberration," but the result of a broader set of problems related to police culture and procedure:

> Despite the outpouring of righteous and anxious statements from university administrators and other public officials disavowing this allegedly isolated instance of excessive state violence, there was a reason why those riot-geared UC Davis officers—who had to know they were performing for the cameras—did not hesitate to (nonlethally) fire. It is not difficult to see that in the post-1960's period, militarized police repression of actual and potential political disorder is as American as the atom bomb. However, what is too quickly taken for granted is that the primary places in which such police power is exercised are the very same that are least likely to send their young people to places like UC Davis, and which are the primary sites for which the state has trained its police to engage in domestic, low-intensity warfare by way of drug sweeps, no-knock warrants, street harassment, traffic stops, and justified use of (deadly) force. What is to be made of the UC Davis police action of November 18, 2011, once we understand that the foundation of those officers' presumed right to violence is not in some failure of bureaucratic protocol, nor in the obvious evidence of poor administrative leadership, but instead derives from the generalized legal, political, and cultural mandate of the US policing apparatus to dispense force as it sees fit?

The use of pepper spray at UC–Davis should not be viewed as an isolated incident. It must be understood in the context of the militarization of the police and the resulting normalization of police violence. As Rodriguez points out, this violence is disproportionately inflicted on people of color and poor people and largely ignored by those outside these communities.

The Militarization of the Police Comes to Campus

The use of pepper spray on peaceful, seated students is but one example of the now common use of force against protesters in the United States. Several other high-profile incidents have recently highlighted this trend in the militarization of American police forces. With no provocation whatsoever, New York police officer Anthony Bologna pepper sprayed a group of women who had already been detained and corralled by police at an Occupy Wall Street protest in September 2011. Two of the women are now suing Bologna and the city. In Oakland former Marine Scott Olsen returned from two tours in Iraq to be struck in the head with a police tear-gas canister during the general strike organized by Occupy Oakland on November 2nd. The police continued to shoot flash bang grenades at protesters as they tried to pull Olsen to safety. He suffered a fractured skull. In November, police at an Occupy protest in Seattle pepper sprayed an 84-year-old retired school teacher while she was penned in by officers on motorcycles. These events and many others like them highlight a pattern emerging throughout the United States. What is behind this repeated use of excessive force against peaceful civilians? Many blame the dramatic rise on the militarization of American police forces over the past few decades.

While militarized police tactics have only recently been spotlighted in media coverage of Occupy protests, they have been used against American citizens for decades, particularly in poor communities of color. Paramilitary-style policing in the United States began in Los Angeles in the 1960s in response to growing civil unrest and rioting in the inner city. These new tactics were steadily adopted by police forces across the country over the next few decades. The "wars" on drugs and terror have provided police with additional justification for militarizing. In the 1980s, Ronald Reagan's anti-drug crusade created more funding and training for paramilitary-style policing, which was largely utilized for routine drug busts in poor communities of color (Balko, 2006). The increasing military capacity of civilian police departments spiked again after the terrorist attacks of September 11, 2001. The government began to provide military equipment and training to departments of any size including small rural towns, some with populations below 5,000 (Stamper, 2011). There are now an estimated 40,000 paramilitary police raids each year in the United States (Balko, 2006). As of 1997, 90 percent of cities with populations of 50,000 or more had SWAT teams (Kraska and Kappeler, 1997). The San Francisco Bay Area in California even has a separate SWAT team tasked with guarding the subway system (Balko, 2006).

None of this is surprising given the generous funding provided by the U.S. government to assist police forces in acquiring military equipment and technology. In 1994, the Department of Defense authorized the allocation of military equipment to civilian police departments, and Congress created a program for transferring used military weaponry and equipment to state and local police. In 1997, the Department of Defense created the Law Enforcement Support Program to facilitate the distribution of military equipment to civilian police departments (Balko, 2006). More than 11,000 federal and state police forces from all 50 states participate in the program. Between 1997 and 2000, over $727 million worth of military equipment was transferred to civilian police departments, including 253 aircraft, 7,856 M-16 rifles, 181 grenade launchers, 8,131 bulletproof helmets, and 1,161 pairs of night-vision goggles (Twohey, 2000).

Although the military and the police are beginning to look more and more alike, for much of American history there has been a clear divide between the two. There is very good reason for this separation. As journalist Radley Balko points out,

> The military and the police have two distinctly different tasks. The military's job is to seek out, overpower, and destroy an enemy. Though soldiers attempt to avoid them, collateral casualties are accepted as inevitable. Police, on the other hand, are charged with "keeping the peace," or "to protect and serve." Their job is to protect the rights of the individuals who live in the communities they serve, not to annihilate an enemy. (2006, 15)

While many law enforcement officials seem to have these roles confused and quickly gobbled up generous government funding, retired New Haven police chief Nick Pastore refused to accept the military's hand-me-downs. In 1999 he told the *New York Times* that providing civilian police with paramilitary gear and training "feeds a mind-set that you're not a police officer serving a community, you're a soldier at war. I had some tough-guy cops in my department pushing for bigger and more hardware . . . . They like SWAT because it's an adventure" (Egan, 1999).

Pastore is not the only law enforcement official to question the militarization of the police. Norm Stamper was chief of the Seattle Police Department in 1999. He later expressed regret about the decisions he made during the infamous WTO protests, particularly the use of chemical agents. In an interview with *Democracy Now*, he stated, "I will regret that decision for the rest of my life. We took a military response to a situation that was fundamentally nonviolent, in which Americans were expressing their views and their values, and used tear gas on them. And that was just plain wrong" (Goodman and Gonzales, 2009). In a 2011 article for *The Nation*, Stamper writes, "Everyday policing is characterized by a SWAT mentality, every other 911 call a military mission. What emerges is a picture of a vital public-safety institution perpetually at war with its own people" (8). Balko points out that the American people are essentially footing the bill for the military-style policing of their communities and the accompanying change in the role and attitude of their police officers. He argues, "The Pentagon used its defense budget to buy that equipment, a budget given to it by Congress on behalf of American taxpayers for the purpose of defending Americans from threats from abroad. It's perverse to then use that equipment against American citizens" (Balko, 2006, 40).

The UC–Davis pepper-spray incident was a wake-up call for many in the local community that college and university campuses are not immune to these changes in police culture and tactics. Sloan argues in a 1992 study that the "modern campus police" developed in the 1960s as a response to the antiwar and civil-rights protests that erupted on campuses across the country. This was around the same time that their civilian counterparts were first using military tactics and technology in Los Angeles. Sloan notes that this new style of campus policing "developed a striking similarity to urban police departments in both their administrative structure and their operation. Their role on campus expanded from custodians to full-fledged law enforcers" (1992, 88). As state and local forces have increasingly adopted military tactics and technology, campus police have followed suit. On UC campuses, paramilitary-style police from neighboring counties have frequently been called in to work crowd control and disperse protests. For example, Davis campus newspaper *The Aggie* reported that 120 police officers from over ten law enforcement agencies were called in to confront protesters as they attempted to march onto the highway on March 4th, 2010 (Steussy and Kumar, 2010).

Conclusion

The problem of assigning responsibility for the pepper-spray incident mirrors the problem faced by student activists more broadly: who is responsible for the privatization and marketization of the university, and what can advocates of public higher education do to stop it? It is important to remember that the reason the students were on the quad protesting on November 18th was not only a response to police brutality towards political activists, but also to speak out against the privatization of the university and economic and racial inequality to which that is linked. The connection between the privatization of once public goods and services and the militarization of the police is often invisible. Instances like the UC–Davis pepper-spray incident make it clear that the role of the police has become more about protecting private property and corporate interests than about protecting citizens. Many activists across the UCs and at Occupy events nationwide are keenly aware of this, as evidenced by the common chant that is often directed at police: "Who do you serve? Who do you protect?" They know it is not the people who are beaten and pepper sprayed as they attempt to peacefully exercise their right to protest, but rather the powerful entities these protests are targeting.

Bibliography

Adejunmobi, M., Biltekoff, C., Carroll, P., de la Cadena, M., Dumit, J., Grindstaff, L., . . . Zhang, L. (2011, November 22). Open letter to the UC Davis Students to explain to them why we are not signing the petition for Chancellor Linda Katehi's resignation. http://occupyucdavis.org/2011/11/open-letter-to-the-uc-davis-students-to-explain-to-them-why-we-are-not-signing-the-petition-for-chancellor-linda-katehi%E2%80%99s-resignation/

AggieTV. (2011, November 22). Thousands protest violence & tuition at UC Davis [Video file]. Retrieved from http://www.youtube.com/watch?v=7c3NiAxhGX4&feature=plcp

Antonelli, G. A., Aranovich, R., Benware, W. A., Bradac, M., Carlip, S., Carroll, P., . . . Ziser, M. (2011a). Endorsers. https://sites.google.com/site/realfactsheet/endorsers.pdf

Antonelli, G. A., Aranovich, R., Benware, W. A., Bradac, M., Carlip, S., Carroll, P., . . . Ziser, M. (2011b). Real UC Davis fact sheet. https://sites.google.com/site/realfactsheet/

realfactsheet.pdf

Balko, R. (2006). *Overkill: The rise of paramilitary police raids in America.* Washington, DC: Cato Institute.

Bicycle Barricade. (2011a, December 2). Task farce. http://bicyclebarricade.wordpress.com/2011/12/02/task-farce/

——. (2011b, December 7). Faith healing: The therapeutic rhetoric of the UC Davis administration. http://bicyclebarricade.wordpress.com/2011/12/07/faith-healing-the-therapeutic-rhetoric-of-the-uc-davis-administration/

——. (2012, February 21). The neoliberal university, banks, and the professors who love them. http://bicyclebarricade.wordpress.com/2012/02/21/the-neoliberal-university-banks-and-the-professors-who-love-them/

Birgeneau, R. J. (2011, November 10). Message to the campus community about "Occupy Cal." http://newscenter.berkeley.edu/2011/11/10/message-to-the-campus-community-about-occupy-cal/

Branan, B. (2012, February 18). UC Davis votes for chancellor's leadership. *Boston Herald.* http://www.bostonherald.com/news/national/west/view/20120217uc_davis_votes_for_chancellors_leadership

Brown, N. (2011, November 18). Open letter to Chancellor Linda P. B. Katehi. http://bicyclebarricade.wordpress.com/2011/11/19/open-letter-to-chancellor-linda-p-b-katehi/

Buscho, D. (2011). Police pepper-spray peaceful UC Davis students: Ask Chancellor Katehi to resign! www.change.org/petitions/police-pepper-spray-peaceful-uc-davis-students-ask-chancellor-katehi-to-resign

California Scholars for Academic Freedom. (2011, November 23). Letter to Yudof and Regents from California Scholars for Academic Freedom. reclaimuc.blogspot.com/2011/11/letter-to-yudof-and-regents-from.html

Campus protests: Continuing the dialogue. (2012, Winter). *UC Davis Magazine, 29*(2), 2–3.

Carolina, B. (2011). Denounce calls for Chancellor Katehi's resignation at UC Davis: Support Dr. Katehi as the UC Davis chancellor. http://www.change.org/petitions/denounce-calls-for-chancellor-katehis-resignation-at-uc-davis-support-dr-katehi-as-the-uc-davis-chancellor

CBS Sacramento. (2012, January 27). UCD paid $100,000 for media consultant after pepper spray incident. http://sacramento.cbslocal.com/2012/01/27/ucd-paid-100000-for-media-consultant-after-pepper-spray-incident/

Clover, J., Ferguson, M., Wittman, D., and Cox, D. (2011). The faculty, the fact sheet, and the facts: An honest accounting of November 18. http://ucdfa.org/wp-content/uploads/2011/12/facultyfactsheet.pdf

Davis Faculty Association. (2011a, November 19). DFA Board calls for Katehi's resignation. http://ucdfa.org/2011/11/19/dfa-board-calls-for-katehis-resignation

——. (2011b, December 4). Other groups call for Katehi to resign. http://ucdfa.org/2011/12/04/other-groups-call-for-katehi-to-resign/

——. (2011c, December 11). Member survey results. http://ucdfa.org/2011/12/11/member-survey-results/

Egan,T. (1999, March 1). Soldiers of the drug war remain on duty. *New York Times.* http://www.nytimes.com/1999/03/01/us/soldiers-of-the-drug-war-remain-on-duty.html

Filler, D. (2011, December 2). Policing's "new normal" doesn't work with white folks. *Davis Enterprise.* http://www.davisenterprise.com/opinion/opinion-columns/policings-new-normal-doesnt-work-with-white-folks/

Ganascia, J.-G. (2010). The generalized sousveillance society. *Social Science Information, 49*(3), 1–19.

Gaynor, S. (2011, December 2). UC's pick of Bratton to investigate pepper spray incident isn't reassuring. *San Francisco Bay Guardian.* http://www.sfbg.com/politics/2011/12/02/ucs-pick-bratton-investigate-pepper-spray-incident-isnt-reassuring

Golden, C. (2011a, November 21). UCD police chief placed on administrative leave. *Davis Enterprise.* http://www.davisenterprise.com/local-news/crime-fire-courts/protests-again-gathering-steam-on-campus/

——. (2011b, November 22) Former LAPD chief to head pepper-spraying probe. *Davis Enterprise.* http://www.davisenterprise.com/local-news/crime-fire-courts/former-lapd-chief-to-head-pepper-spraying-probe/

——. (2011c, December 3). Katehi, faculty agree on message, not who can deliver it. *Davis Enterprise.* http://www.davisenterprise.com/local-news/ucd/katehi-faculty-agree-on-message-not-who-can-deliver-it/

——. (2012a, January 26). UC spent $100,000 on consultant after pepper-spraying. *Davis Enterprise.* http://www.davisenterprise.com/local-news/crime-fire-courts/uc-spent-100000-on-consultant-after-pepper-spraying

——. (2012b, February 16). Pepper spray report delayed again. *Davis Enterprise.* http://www.davisenterprise.com/local-news/crime-fire-courts/pepper-spray-report-delayed-again

Goodman, A., and Gonzalez, J. (2009, March 30). "I made major mistakes"—ex-Seattle police chief admits response to 1999 WTO protests was too heavy-handed. *Democracy Now.* http://www.democracynow.org/2009/3/30/i_made_major_mistakes_ex_seattle

Grindstaff, L. (2012, January 27). Who has access to what? http://ucfacultysupportingstudents.org/2012/01/27/who-has-access-to-what/

Harper, B. (2011, November 23). Occupy white supremacy: What UC Davis pepper spraying tells me about the racialized politics of sentimentality. http://sistahvegan.wordpress.com/2011/11/23/occupy-white-supremacy-what-uc-davis-pepper-spraying-tells-me-about-the-racialized-politics-of-sentimentality/

Katehi, L. P. B. (2011, November 18). Chancellor addresses Friday's removal of tents from the Quad. http://chancellor.ucdavis.edu/messages/2011/protest_action_111811.html

——. (2012, Winter). A message from the chancellor. *UC Davis Magazine, 29*(2).

KCRA TV. (2011, November 20). Pepper spray officers placed on leave. http://www.youtube.com/watch?v=nx5xGR_Bopg

King, J. (2011, August 17). The US is still recovering from "supercop" Bill Bratton's policing. *Colorlines.* http://colorlines.com/archives/2011/08/will_supercop_bill_brattons_zero_tolerance_work_in_the_uk_us_no_-_heres_why.html

Kraska, P. B., and Kappeler, V. E. (1997). Militarizing American police: The rise and normalization of paramilitary units. *Social Problems, 13*, 1–18.

Leal, W. S., Amenta, N., Samaniego, F. J., Castro, R. H. R., Johnson, K., Blumwald, E., . . . Shoemaker, C. F. (2011, November 30). Katehi has faculty support [Letter to the editor]. *Davis Enterprise.* http://www.davisenterprise.com/opinion/letters/katehi-has-faculty-support/

Mama, A., Craig, M., Constable, L., Ho, W., Joseph, S., Kaiser, S., . . . Swain, M. (2011, November 20). Open letter to the chancellor. http://bmcdb.wordpress.com/2011/11/21/another-open-letter-to-chancellor-katehi-from-women-and-gender-studies-faculty/

McCarthy, B., Jackman, M., U, E., Wolf, D., Smith, V., Hall, J., . . Chiaraluce, C. (2011). Open letter to Katehi from Sociology Department members. https://docs.google.com/

open?id=0B7h6xM-Rj76RMjgzMTMzOTYtMWNjOS00YWJhLTlhOTItZTdiNDE3YmM5MGUz

Meister, R. (2011, November 27). Letter to President Yudof objecting to hiring William Bratton to investigate UC Davis pepper spray incident. http://cucfa.org/news/2011_nov27.php

Meranze, M. (2012, February 19). Davis Agonistes. http://utotherescue.blogspot.com/2012/02/davis-agonistes.html

Ostertag, B. (2011a, November 19). Militarization of campus police. *Huffington Post*. http://www.huffingtonpost.com/bob-ostertag/uc-davis-protest_b_1103039.html

——. (2011b, December 21). The chancellor and the student, or, speaking truth to power. *Huffington Post*. http://www.huffingtonpost.com/bob-ostertag/uc-davis-pepper-spray_b_1161409.html

Price, A. (2011, November 30). Is UC Davis's pepper spray investigation truly independent? *The Atlantic*. http://www.theatlantic.com/national/archive/2011/11/is-uc-daviss-pepper-spray-investigation-truly-independent/249263/

Pringle, P., Quinones, S., and Grad, S. (2011, November 20). UC Davis police defend use of pepper spray on Occupy protesters. http://latimesblogs.latimes.com/lanow/2011/11/uc-davis-police-defend-use-of-pepper-spray-on-protesters.html

Rodriguez, D. (2011, December 9). De-provincializing police violence: On the recent events at UC Davis. http://reclaimuc.blogspot.com/2011/12/de-provincializing-police-violence-on.html

Sakash, T. (2011, November 23). Students press UCD administrators on police actions. *Davis Enterprise*. http://www.davisenterprise.com/local-news/crime-fire-courts/students-press-ucd-administrators-on-police-actions/

Setele, J., and Grindstaff, L. (2011). Making sense of (in)civility: Voices from UC Davis. http://civilityproject.ucdavis.edu/makingsense/

Sloan, J. (1992). The modern campus police: An analysis of their evolution, structure, and function. *American Journal of Police*, *11*(2), 85–104.

Stamper, N. (2011, November 28). Occupy the police. *The Nation*, 6–8.

Stanton, S. (2011, November 23). Katehi: Campus police were told not to use force on students. *Sacramento Bee*. http://www.sacbee.com/2011/11/23/4074993/katehi-campus-police-were-told.html

Steussy, L., and Kumar, P. (2010, March 5). Protestors clash with police in attempt to block I-80. *The Aggie*. http://www.theaggie.org/2010/03/05/protestors-clash-with-police-in-attempt-to-block-i80/

Twohey, M. (2000, January 1). SWATs under fire. *National Journal*, 37.

UC Davis Chancellor's Office. (2011, November 29). Fact sheet on recent campus demonstrations at UC Davis. http://chancellor.ucdavis.edu/local_resources/pdfs/20111129_Fact%20Sheet-FINAL_crx.pdf

UC Davis Dateline. (2011a, November 21). Chancellor apologizes for pepper spraying. http://dateline.ucdavis.edu/dl_detail.php?id=13761

——. (2011b, November 23). Town hall: Chancellor, others say "never again." http://dateline.ucdavis.edu/dl_detail.php?id=13764

——. (2011c, November 28). Emeritus Cruz Reynoso to lead task force on pepper spraying. http://dateline.ucdavis.edu/dl_detail.php?id=13767

UC Davis Facts. (2011). Annual fees and tuition, full-time attendance. http://facts.ucdavis.edu/annual_fees_and_tuition.lasso

UCLA Faculty Association. (2011, December 26). UC excerpt from Dec. 14 legislative pepper spray hearing. http://uclafacultyassociation.blogspot.com/2011/12/uc-excerpt-from-dec-14-legislative.html

Weitzer, R., and Tuch, S. A. (2004). Race and perceptions of police misconduct. *Social Problems, 51*(3), 305–325.

Wittman, D. M. (2012, January 24). How I got involved. http://ucfacultysupportingstudents.org/2012/01/24/how-i-got-involved/

• CHAPTER TWELVE •

Higher Ed on a Slippery Slope

Pulling It Back from the Brink of Tyranny

Maura Stephens

Six months into the Occupy movement, Goddard College president Barbara Vacarr hosted "Occupy Goddard," the first academic conference devoted to the phenomenon of peaceful populist resistance to oppression and economic inequities that has been sweeping the nation (and beyond). The conference brought together local activists with those from far afield for a day of panels and participation, culminating in the first statewide Vermont General Assembly.

It was a decent step, but as of twelve months later, it was disappointing (if not surprising) to realize that nothing discernible had come of it outside a few good activist connections made that day. Higher education ought to be playing a central part in the Occupy and related movements, but except for a few activist campuses, higher ed has been relatively invisible—except when itself engages in exactly what these movements are protesting: cracking down on free speech, academic freedom, and the whole time-honored idea of liberal education of critical thinkers unfettered by corporate ties.

And, appallingly, it *is* engaging in such antidemocratic activities, to an alarming degree.

Especially in comparison to Burma, Egypt, Iran, and other repressive regimes, people in the United States have until recently enjoyed increasing civil liberties and rights. These rights and liberties were often hard won by those who went before us, risking much to secure them for the good of all.

By the time we get to college, most of us have at least a smattering of knowledge of some historic peaceful people's struggles—emancipation, civil rights, Gandhi and Indian independence, ending apartheid—but we haven't necessarily engaged in peaceful actions ourselves, because most of these struggles seem long ago and far away.

But this is not our parents'—or even our own—USA. Across the country, both private and public universities are being corporatized, as evidenced by spiking tuitions; the crushing of faculty and staff unions; the ballooning of the use of part-time, temporary full-time, and adjunct faculty (now comprising 70 percent of all university and college instructors, they receive paltry wages for their hard work, besides being treated as third-class employees without benefits or job security); the unapologetic repression of freedom of speech and traditional academic freedom; the surge in vocational education for job and career rather than for citizenship and reasoned thinking; and the rise of for-profit colleges.

On campus and off, we are witnessing an increasing military-police crackdown. Peaceful protesters have been harassed with rising regularity by campus police, regular police, FBI, and "Homeland Security"—and perhaps by police-for-hire as well (it's sometimes hard to tell).

Wake Up and Smell the Tear Gas

Soon after news of the UC–Davis police brutality broke (see below), I sent a message (bcc) to several dozen full-time faculty members on my own campus regarding making a public statement against intimidation tactics. (I wouldn't put a part-time faculty member in the position of having to choose between salary, minimal as it is, and conscience.) I received two responses, one from a colleague ready to join forces and one a tepid "maybe."

In their defense, my colleagues' workloads are heavy, many of them have large student-loan repayments, at least in their first decade or so on the job, and they are pressured not only to teach but to perform community service and to conduct and publish research. These stresses come on top of their own life obligations. (On the other hand, they generally don't have to work two or three jobs to keep food on the table, on top of their own life obligations, and they get paid vacations and health and retirement benefits.)

Perhaps they've just been too busy to be aware of the intimidation tactics and violence being unleashed against ordinary people, including students, who are engaged in protesting injustice in various forms.

But it's past time for concerned citizens and campus communities to get out front and take a public stand against harassment, intimidation, and violence on campuses, especially when it is aimed at peaceful protesters, whether they are campus based or not.

So this is for all people who work in academia around the country, from pre-school on up: If you have been too busy, you'd better take some time to catch up. Your career, indeed your very future, may depend on it.

Unhealthy, Undemocratic Crackdown

Here are just a few examples of a sweeping and terrifying anti-intellectual, antidemocratic trend on campuses in the United States—land of the free and home of the brave, far from Belarus and Burma, Rwanda and Syria, Tunisia and Iran:

UC–Berkeley

The University of California–Berkeley's Sproul Plaza almost echoes with history. It was home to the 1964 Free Speech Movement and later protests against the war on Vietnam, apartheid, wars on Iraq, and other atrocities. During the fall semester 2011, fittingly, it became the home of an Occupy encampment.

On November 9, the plaza was the setting for a campus-wide day of action supporting affordable higher education and the Occupy movement, and protesting state budget cuts that harm access to education. Activists staged "teach-outs," a noontime rally, and a march to the Bank of America on Telegraph Avenue.

But things got ugly as UC–Berkeley's police began beating peaceful students with batons, even as students chanted "Don't beat students! Don't beat students!" and "Your families will see this!" Later, police tried to clear the Occupy encampment, and students formed a human barricade around their tents. The police shoved students aside, beating many. Thirty-nine students and a professor were arrested, charged with failing to disperse and resisting arrest. Many participants webcast videos from the scene, which were widely viewed.

UC–Davis

Nine days later, on November 18, another UC campus, Davis, was the scene of more police brutality against peacefully protesting students. Sent in to use force to clear the Occupy Wall Street UC–Davis encampment, police officers in riot gear found that most protesters had already cleared their tents. The police moved out the few remaining tents, and then students formed a wide-seated circle around the officers, not impeding the officers' movement but making a statement.

Before officers could even try to negotiate with students, one of the officers, Lt. John Pike, deliberately and heavily pepper sprayed seated students who posed no threat. One of the victims was sent to the hospital to be treated for chemical burns. The seated students, incapacitated now, were attacked by the police with batons and other physical force, yet still remained peaceful. A video of the incident has been viewed, as of this writing, more than 2.5 million times.

In a statement, Bernie Goldsmith of the nearby Occupy Davis (separate from Occupy UC Davis), wrote: "At Occupy Davis, relations with the democratically elected city council and local police forces have been genial and productive. . . . Those in charge of using force are aware that they are democratically elected officials that are directly accountable to the people. Occupy UC Davis, a mere three blocks away, is under the jurisdiction of an undemocratic, appointed regime of force over which its subjects have no meaningful democratic control. The authorities there attacked non-violent protesters with indifference, and, in some cases, a clear display of sadistic pleasure" (Cherkis, 2011).

The university administration admitted, "There was no student violence by any protesters."

On November 20, someone who identified as Wat Tyler (presumably a pseudonym, after the 14th-century English revolutionary who led the English Peasants' Revolt of 1381) wrote this on a blog about the UC–Davis actions:

> I saw the whole thing, from start to finish. I saw a group of heavily armed police in riot gear arrive, under instructions from the university chancellor, to remove a small group of peaceful student protesters who had set up camp of seven tents on a tiny section of a wide open space on the campus of a public university. She ordered this on spurious health and safety grounds, a rationale so laughable that very quickly she changed her story and introduced instead a bizarre tale of non-UCD agitators infiltrating the camp. Again, this is ridiculous. I know many of the protesters personally and all are either current UCD students, instructors, or alumni. (Student Activism, 2011)
>
> The police moved in violently from the outset and to everybody who was present it was clear that they had arrived specifically to intimidate with force. Whether they were right or not to arrest the protesters for obstructing a piece of grass on a huge expanse of grass is actually not even an issue. Because a university is by its very nature a forum for the peaceful and respectful exchange of ideas—and the UCD quad itself was designed with that very principle in mind—the police simply should never have been on campus in the first place, and they most certainly should not have been dressed in paramilitary battle gear. But above all of this they absolutely should not have assaulted the students with a military grade chemical agent at point blank range.
>
> Ultimately, this was a political action by the chancellor that mirrors the actions of other UC chancellors in the past two weeks in seeking to violently silence dissent as UC faculty, staff and students mobilize to defend our public university against privatization moves initiated by the plutocracy that is the UC Board of Regents. . . . It really is as simple as it appears. And that is why today the UCD general assembly of 5,000 students, staff, faculty, alumni, parents, and local citizens met on that very same quad and voted with a majority of 95% to hold a general strike next week to demand immediate reforms and changes to the way the university is governed and policed.

In both the UC–Berkeley and UC–Davis incidents, the police violence was unprovoked and quite clearly in violation of the University of California's "Universitywide Police Policies and Administrative Procedures," which states, "Chemical agents are weapons used to minimize the potential for injury to officers, offenders, or other persons. They should only be used in situations where such force reasonably appears justified and necessary"; and "No officer shall strike an arrestee or suspect except in self-defense, to prevent an escape, or to prevent injury to another person" (Student Activism, 2011).

CUNY

The campus violence didn't stop in California. Three days after the UC–Davis brutality, students at City University of New York Baruch College were protesting tuition hikes, unfair labor practices targeted at adjunct and other faculty, and the privatization of the CUNY system. When the students attempted to enter an auditorium where the trustees were holding a public hearing, CUNY police officers pushed them back with batons; 15 students were arrested, and a police officer was apparently injured, although details were not released on the officer's injury.

Indiana University

The violence isn't a coastal phenomenon, either. On November 29, 2011, five students at Indiana University Bloomington were arrested for peacefully protesting überbank JP Morgan Chase, which had come to recruit at the Kelley School of Business. Chanting "Not on our campus! Not on our campus!" other students were shoved and assaulted, especially by one IUB detective.

"JP Morgan-Chase was among the major financial institutions that caused the 2008 financial collapse with its criminally greedy, fraudulent lending practices (in the forms of mortgage-backed securities and collateralized debt obligations)" wrote the students in a press release (Leonard, 2011). "It should have failed then, but because of the influence Wall Street has over the government, it was bailed out, . . . has not ceased in . . . fraudulent practices, . . . contributed enormously to the corruption of our democracy, has caused misery on a massive, debilitating scale within this country, and has a list of other crimes to its name that is too lengthy to enumerate."

Waging War on Higher Ed's Purpose

All of these protests indicate that the student activists were engaging in the kind of critical thinking we are supposedly encouraging in college students. They recognized that our government is spending money on things unrelated to the furtherance of uncorporatized education. They see that universities are as culpable as private corporations in trampling worker rights. They connect the dots between banks' corrupt business practices, the government's collusion, and the harm caused to individuals. This is just smart. This expression of students' rage against the machine is something to applaud, not abort.

There's more to the assault on higher education than "just" tuition hikes, labor repression, squelching free speech and assembly, collusion with corporations, and violence against peaceful protesters.

Even with all this, and despite nationwide public outrage at mass shootings at Virginia Tech in 2007 (32 students and faculty killed, 15 wounded) and Northern Illinois University in 2008 (five students killed, 16 wounded), the elementary school in Newtown, Connecticut in 2012 (20 children, six adults killed), and the too-many-to-count shootings at high schools since a sharp rise in their occurrence starting in the late 1980s, the extremist pro-gun lobby is pushing legislation that would prohibit colleges and universities from banning or regulating firearms on campus and would allow or even encourage students to own and carry concealed handguns. This bizarre and frightening movement has succeeded to the extent that at least 25 two- and four-year schools, in Colorado, Michigan, Utah, and Virginia, now allow the carrying of firearms on their premises.

On December 31, 2011, President Barack Obama signed into law the National Defense Authorization Act (NDAA), a terrifying piece of legislation if ever there was one. It authorizes indefinite military detention of anyone, including U.S. citizens, picked up anywhere in the world, based on the (any) president's suspicion that that person was engaged in "subversive" activities—which could include peaceful protest on college campuses or anywhere else—and could be indefinitely detained without charge or access to counsel or human contact, "disappeared," or killed. Legally. (The ACLU has a toolkit to help fight back against this with model state and local legislation and other resources.)

In early March 2012, President Obama signed into law the Federal Restricted Buildings and Grounds Improvement Act of (H.R. 347, also known as the "Trespass" or "anti-protester" or "anti-Occupy" bill). This law rewrote a 1971 trespass law. Now it's an offense to knowingly enter certain areas without "legal authority." These areas include the White House or the VP's official residence and grounds; a building where the president or any other person protected by the Secret Service is or will be visiting (many people nowadays are protected by Secret Service, and visits are not always announced publicly in advance); or a building or grounds restricted "due to a special event of national significance" (which can be very loosely defined).

In late April 2012, Seattle, Washington, became the first U.S. city to receive Federal Aviation Administration approval to use two unmanned aerial drones it had purchased. When the public became alarmed, the city's police chief assured them that the drones would not be used "until written policies are in place for their use." Small comfort to those who have been harassed, intimidated, and attacked on the ground to think they will be spied upon by "eyes in the sky" henceforth.

What Can Campuses Do?

This may seem all very new and scary, and to the degree and speed with which it is overtaking our universities, it sure is. But if we hide, it will not go away. If we are silent, we are complicit in our own repression. As scholars and thinkers, as people who are expected to lead, as people who are ostensibly teaching our students to be critical thinkers who question authority—especially abusive authority—we have a moral obligation to speak out when we see injustice and subversion of humane and democratic ideals on which our society and nation claim to be based.

So, what should faculty, staff, students, and administrators with integrity do?

1. Hold an open series of dialogues about your individual roles and your campus's role in protecting and fostering peaceful protests such as those of the various Occupy movements across the country (and the world). Invite regular Occupy participants and theorists from your region. Listen.

2. Get on a bus or train, carpool, bike, or walk to the closest Occupy and spend some time listening to people. Be sure to time it so you witness a General Assembly. Watch respectful democracy in action; many of us—and probably most students—have never actually witnessed such a process, except perhaps within our own families or small groups.

3. Stop watching Fox "News" and other conglomerate-run news media. Switch to independent, non-corporate-run news sources. Watch, listen to, or read the daily transcripts of *Democracy Now!*, watch Free Speech TV and The Young Turks, listen to Free Speech Radio News, read Alternet, Common Dreams, TruthDig, Truthout, and other independent media. (Full disclosure: I write for these publications.)

Visit and read occupywallstreet.com. Find your local Occupy and engage. Help stop foreclosures, move money into local credit unions, strengthen labor, fight tuition hikes, support local farmers, ban fracking, and related issues.

4. Learn about what issues are important in your university or college community, your hometown, your state, your nation. Find your personal passions for peace, rights of Nature, environmental sensibility, human rights and equity, electoral and economic justice, animal rights. Become engaged in or start an Occupy or Occupy-inspired movement on your campus. Learn the ropes through your participation with an established Occupy group.

5. As faculty and/or staff, cowrite and co-edit a statement of solidarity with those who exercise their civil liberties and rights—of free speech and free assembly, to protest injustice and immoral laws, to academic freedom, to have unfettered professional and citizen media, to not be attacked while peacefully demonstrating one's resistance to corporate or state tyranny. Point the finger at those on your campus or in your community, state, or nation who are trampling those rights and liberties, and demand action to stop that co-option.

6. Get your administration and trustees to discuss peaceful protest and urge them to make a statement of solidarity as well.

7. Take a careful look at the trustees of your institution. Were they chosen for their intellect, creativity, commitment to higher education, critical thinking skills, expertise in an important discipline, and passion for learning, or were they chosen for their big bucks and corporate connections? What institutions, other than yours, do they represent? Where do they earn their living? Do they represent you (and the rest of the 99 percent)? Does your institution's governing board include representatives of staff, faculty, student, alumni, and neighbor constituencies? Are they given equal say on committees and other governing mechanisms of the institution? If not, why not?

8. What are buildings and facilities on your campus named after? Are corporations being honored? If so, there's too much of a corporate financial tie to your university. Where does research money come from, if you're at a research institution? Is it all Bank of America, Chevron, Citigroup, Dow, ExxonMobil, GlaxoSmithKline, Goldman Sachs, Halliburton, Monsanto, Raytheon, Wal-Mart, and their ilk? You might want to dig deeper. If the university has got itself in too deep with industry, see if you can find student, citizen, or professional journalists to investigate the relationships. Or, rather than trying futilely to reform it from within, you might look elsewhere for work (easier said than done, I realize) or education and then get the word out as widely as possible about your former institution's ties.

9. Take a careful look at the endowment of your institution. Where is it invested? Take a look at your own pension plan (TIAA-CREF or other). Where is it invested? Are the stocks and bonds and real estate holdings in companies or countries that are exploiting workers, polluting the air and water, oppressing residents, pillaging nature, privatizing the commons? Start a campaign to move these investments into more humane and sustainable enterprises. Work with the chief financial officer of the institution and the board of trustees to help them understand how powerful movements for change can be heightened by the investment choices of higher-ed institutions. Together, colleges

and universities are some of the biggest private investor blocs in the world. Visit endowmentethics.org to jump-start a movement to divest your institution from fossil fuels.

10. Do all the employees of your institution receive a living wage (not just minimum wage) for your community? Do they all receive decent health benefits and retirement savings opportunities? Do adjunct/part-time faculty members receive fair compensation? Do all employees receive educational benefits and/or professional development opportunities? Is there a fair employment policy in general, with an effective and equitable system for redressing grievances? Is the institution supportive of unions? What is its history with labor? Has it ever busted a union or kept one from forming? What is its stated policy toward labor? Does it negotiate fairly, if it does have unions?

If the institution comes up short on these issues, it's incumbent on the best compensated, most well-protected members of the campus community to change all these things—that, generally speaking, means the tenured faculty have the responsibility to speak out on behalf of not just themselves but all those who don't have the job guarantee they enjoy.

Further, there are probably people on your campus who have lost their homes to foreclosure or inability to pay the rent. What kind of community sharing program exists at your institution, if any? If none, start one. Get the chaplains, the chief financial officer, the development or fundraising office, and "human resources" to establish and contribute to an emergency relief fund and fundraise from your fellow employees and the greater community. Occupy a foreclosed home as the Occupy Our Homes movement is doing in cities from Tampa to Chicago to Seattle.

11. Look into the vendors hired by your institution. Even if your university itself doesn't engage in unfair labor practices, its subcontractors might. Does the food service company also run prisons? Does it buy food or supplies from companies that use genetically engineered ("modified") foods ("genfoods")? From agribusinesses that abuse animals? From seafood suppliers that don't label region of origin? If those practices are repugnant to you, you might start a movement to pressure the college to get rid of that vendor or insist that no genfoods be foisted upon the campus community, unless they are clearly labeled as such at the point of purchase or serving. Does the uniform supplier also supply uniforms to private prisons, which run to make money and are not as accountable to the public as are government-run prisons? And so forth. Look around and see where your tuition dollars are going.

12. Get your university to sign onto the KeepGunsOffCampus.org resolution opposing any legislation that would preempt the institution's right to prohibit or adopt policies to regulate possession of firearms on campus.

13. Does your institution have uncensored student-run print and online media, TV, and radio stations? Do they practice actual journalism or PR/marketing? Is your institution's alumni magazine full of feel-good fluff pieces or substantive pieces that address important social and institutional issues with a critical eye? If the answers point to censorship or trickery, do something to change this. Start a website for employees to address grievances and follow through with the administration. Or start a weekly hour-long news and talk show that focuses on issues being addressed by Occupy, as well as labor and other issues that pertain specifically to your campus and its host community.

Talk to the journalism faculty, if there is one, and ask them to start a Citizen Journalism or Community Journalism course and eventually a program for nonmajors. Urge them to push it through. It can't take the usual geologic academia time to introduce this and get it approved.

14. Does the institution live up to its mission statement and other public claims? Does it purport to be inclusive of diversity yet not actively engage in recruiting culturally diverse faculty, staff, and students? Does it claim to be all about sustainability yet leave lights blazing in every locked building with administrators' computers left on day and night? How can you help change the culture so that the mission is foremost in people's minds, or how can you change the mission to reflect and expose the institution's less lofty truth?

15. Enlist the aid of parents, alumni, and students (future alumni), who are considered by many if not most administrators the prime constituents of an institution because they're the donors or potential donors. Unfortunately, that's still how we are valued (or not valued). That's one thing that won't change unless and until the whole education system is revamped.

Because, sadly, the education system is not supported by our government, and money presently drives all private and public education in the United States.

16. Another way to get your institution to support Occupy and related movements is to pressure for more higher-education-supporting legislation on the state and national levels. Primary, secondary, and higher education must have a strong voice in Washington, D.C., to fight the ultra-rightwing assault on education—an assault so real they want to completely eliminate the Department of Education and have all levels of education, from pre–K to PhD, run by corporations. The free-market system at its worst.

17. In a truly fair system, colleges and universities (and churches, mosques, and synagogues) would be paying their fair share of property taxes to the communities where they are based. Economist, author, and professor emeritus of the University of Massachusetts Amherst Richard Woolf points out that the city of New Haven is the poorest community in Connecticut, yet it houses one of the richest universities in the country (indeed, the world). Yet Yale pays no property taxes on its vast holdings of buildings and real estate. The same is true of the other big universities that use the services provided by their cities and towns—at individual and business taxpayers' expense. That system should be abolished over time, say 10 to 15 years.

18. Organize a series on peaceful people's uprisings, on what's known as Nonviolent Direct Action. To get started, read and discuss the works of Gene Sharp and visit the impressive, if incomplete, database of nonviolent works that was compiled by Swarthmore students with other collaborators in 2011 (nvdatabase.swarthmore.edu).

19. Survey the curricula throughout your institution. Identify courses that already have a strong economic and social-justice component. Ask those faculty members (and undergrad and grad students) to work on a program to infuse those components into all syllabi in all disciplines. This kind of thing is being done already on many campuses around environmental, economic, and social sustainability and around diversity and inclusiveness. These subjects are all interdisciplinary and all intertwined, so involve the faculty and students (and staff) who are already engaged in those related efforts to help boost this one.

Let's Change "Business as Usual"

It's time for those of us who work and study on a college campus to do what does not come naturally: act swiftly and decisively, with conviction, and in solidarity. We all stand to lose. We cannot hide behind scholarship, publication pressures, community service obligations, or teaching loads. We have to lead the way, or at the very least join in the movement to build a more equitable and sustainable economy, culture, and government.

20. On May 1, 2012, a national coalition of Occupy assemblies engaged in a "Day without the 99%" general strike, a day away from school and the workplace, a day away from shopping and banking, to shine a light on the "the way the system has enslaved us and burdened us with unmanageable debt, incredibly long working weeks, unfeasibly expensive healthcare." The idea was that

one's absence from work and school would make one's displeasure with this corrupt system known.

But the numbers weren't great enough to make this action the powerhouse it needs to be. If there is another call for such a general strike, administrators, deans, and chairs should give their blessing and encourage students, faculty, and staff to participate. There are great educational benefits to engaging in something like a general strike or other nonviolent direct action outside the workplace and classroom. And this will help the university in the long run, building the movement for more government funding for education and building students' and employees' civic consciousness and experience.

For all of these actions and activities, take time to reflect on your work, celebrate success, and invite new voices at every step.

Let's Live "Campus Community"

Scholarship can seem lonely sometimes, but on a campus we are inextricably linked with those around us—whether it's the IT person who helps you with email problems; the campus safety people who let you into your car or office when you accidentally lock the keys inside; the administrative assistants who help you with copying and ordering and travel and every other thing; the facilities attendants who clear the snow or empty your trash or clean the bathrooms (do you know their names or anything about them?); the dining services chefs and servers who feed you; the librarians who help you procure classroom and research material; the benefits professionals who do their utmost to find the best and least expensive options for health insurance in a maddening bureaucratic labyrinth of options; the biology professor who leads nature walks and teaches mushroom foraging on his own time; the people who teach yoga and kick-boxing at the gym; the admission recruiters who ensure our classrooms are full of the best and brightest; the mental and physical health providers who keep those students well; the career services advisers who help them once they leave us; and the fundraising folks who engage donors so our paychecks continue to be arrive on time and so programs and services will be protected.

Imagine the power if they all banded together in solidarity. After all, they—we—are all the 99 percent. They can't lock us all up, and they can't stop We the People united.

And then imagine ending up like this (big nod to Martin Niemöller): "First they came for the anarchists, but I'm not an anarchist, don't know any anarchists, and don't really understand them. Then they came for the atheists and agnostics, and I thought, good riddance to bad rubbish. Then they came for the sociologists, and I was quiet because I never liked sociology all that much. Then they came for the physicists, and I don't know anything about physics so I didn't care. And then they came for the musicians, artists, and theater people, and I didn't notice right away until I realized it was getting boring and awfully quiet around here. Then they came for the athletes, and there was nothing left for me to do on the weekends. Then they came for the journalists, and there was nobody left to tell me what was happening. Then they came for the theologians and philosophers, and I had no one left to help me understand. And now I'm sitting in this cell, wondering how much longer I can stand the loneliness—in this nightmare of grief and loss, that's the hardest thing to bear."

Let's get engaged, colleagues. We must. We really, really must.

Bibliography

Cherkis, J. (2011). UC Davis Police Pepper-Spray Seated Students in Occupy Dispute. *Huffington Post*. (November 19). Retrieved from http://www.huffingtonpost.com/2011/11/19/uc-davis-police-pepper-spray-students_n_1102728.htmld

Leonard, M. (2011). IU describes plan to review Occupy protest, arrests at Kelley School of Business. *Bloomington Herald Times*. (December 2). Retrieved from http://newsinfo.iu.edu/news/page/normal/20518.html

Student Activism. (2011) (Blog) "Ten Things You Should Know About Friday's UC Davis Police Violence" (November 22). Retrieved from http://studentactivism.net/2011/11/20/ten-things-you-should-know-about-fridays-uc-davis-police-violence/

CHAPTER THIRTEEN

Occupy Colleges

The Resurgence of Radical U.S. Student Activism

Ryan Thomson and Natalia Abrams

We do not speak on behalf of any participant of the Occupy Colleges Movement. We are not leaders, representatives, managers, directors, or spokespeople; we are members of a movement that leads, represents, manages, directs, and speaks for itself. We are only two core-facilitators and active organizers (like so many others), sharing in a small part of a growing and diverse movement owned by no one. The student movement spontaneously cooperates in its unrelenting struggle to defend public education from the business model's repressive practices of private knowledge production. As the academic-industrial complex becomes increasingly corporatized and factory-like, many of us will resist and serve as the monkey wrenches to break the cogs of the intellectual machine (Nocella, Best, and McLaren, 2010). By nature, students will continue to resist austerity cuts and sabotage conservative corporate agendas because of the felt effects within the classroom. Students, even when they are gunned down, as they were at Kent State University in 1970 or pepper sprayed as they were at the University of California–Davis in 2011, will always fight back and continue to resist academic authoritarianism! In its place, we seek to inspire a participatory education system that overcomes passive learning methods by seeking to facilitate the public's ability to critically reason around a multitude of issues. The millennial generation (born between 1982 and 2000) faces a much larger dilemma beyond that of the broken education system. A deteriorating biosphere, a devaluation of currency, over a trillion dollars in student debt, record inequality, and rampant militarization (both domestically and abroad) constitute a few of the interrelated elements within the grander crisis of the world-system that we will soon inherit.

Symptoms of Educational Crisis

> Our administrators ask of us that we both participate and not participate in our educational system. We are told we must learn to make responsible decisions, yet we are not allowed to make actual decisions. We are told that education is an active process, yet we are passively trained. We are criticized for our apathy and for our activism. In the name of freedom, we are trained to obey. The system requires that we passively agree to be manipulated. But our vision is one of active participation and democratic knowledge production. And this is a demand that our administrators cannot meet without putting themselves out of a job. That is exactly why we should be making the demand. (Davidson *Towards a Student Syndicalist Movement* 1967, p. 43).

Neoliberal education ultimately means corporatization of higher learning and privatization of "charter" K–12 schools that further emphasis tract-based scholarship. This approach devalues students as consumers and teachers as machines rather than depicting both as active participants in the production of knowledge. Students do not need to be told that something is wrong. These conditions are evident both on and off campus. Occupy Colleges cannot take credit for the growing sense of malaise and alienation among the student body. Almost all of us face similar problems: debt, oversize classes, decreased class sections, and casualized professors.

Degree Factories

The university is currently being governed as though it were a juvenile country club run by and to the benefit of big business. In reality, college has become the last party before poverty. Annual cohorts of roughly 1.8 million under- and unemployed youths jump at the opportunities for mobility by enrolling in these overpriced student camps (National Center for Education Statistics at the Institute for Education Science, 2012). Administrators remain driven by a profit motive that standardizes, stratifies, and pacifies its students as obedient debtors. Administrations have unanimously focused on the popularity of their institutions rather than the core benefits of educating the wider populous. Buildings, sports teams, gourmet dining halls, and college life in general have all become more important than the education students receive. The business model views majors and their corresponding departments deemed outside the specialized interests of a university as targets to be absorbed by larger departments, ultimately diminishing the selection offered to students. Liberal-arts colleges and technical universities are slowly co-evolving to specialized training institutions which cater to the market rather than community needs.

Debt

Student loans are at an all-time high in our country with over 36 million people saddled with education debt (Kaplan, 2012). With mass defaults on the rise, student defaulters face steep penalties, the seizure of professional licenses, and the destruction of credit ratings. The cost of higher education has risen over 900 percent since 1972, 650 points above inflation (Harris, 2011). These rising costs have fueled escalating federal and private loans, resulting in a trillion-dollar debt bubble that reduces students and graduates to the status of a new debtor class. Textbooks have exhibited a similar exponential pattern.

Unemployment/Underemployment

A *New York Times* editorial recently placed the underemployment figures for college graduates at over 50 percent, which is higher than the national average (Greenhouse, 2012). In other words, one in two recent graduates is unemployed or underemployed (Yen, 2012). "About 1.5 million, or 53.6 percent, of bachelor's degree-holders under the age of 25 last year were jobless or underemployed, the highest share in at least 11 years" (Yen, 2012). Underemployment figures are incredibly important when discussing college graduates since social mobility and increasing employment opportunities tend to be a primary reason for attending. Why should students attend college when the job market primarily consists of retail or service vocations? This reality is becoming apparent to the student body with each passing semester.

Increased Class Size, Decreased Sections

These trends are coupled with a general decrease in the number of sections being offered to the student body. Universities often attempt to make up the reduction by increasing class sizes. Professors and teaching assistants alike face a difficult decision when "fire hazard"-size classes permit little adventure beyond the standardized bubble-sheet method for class assignments. Discourse and creative professor-student dialogue are foundational to the learning process and must be protected if education is to be more than job training. Professors and lecturers have immense potential to aid in our movement. A number of campus chapters have reported having mutually beneficial relations with professional organizations such as the American Association of University Professors (AAUP).

Casualization of Faculty Appointments/Limit Academic Freedom

Budget cuts have increased the teaching loads of senior faculty members and led most universities to hire teachers as casual laborers on a semester or yearly basis (e.g. TBA enrollment). Distinguished faculty and tenured professors alike are currently viewed as burdens to budgets rather than leaders of intellectual communities. Professors are becoming increasingly dispensable. Since 1975, the number of tenured professors has decreased by 15.2 percent (36.5 to 21.3); similarly, the number of tenure-track professors has decreased by 10.4 percent (20.3 to 9.9), whereas the ratio of full-time non-tenure track faculty members has increased by 5.5 percent (13 to 18.5), and the number of part-time faculty jumped by 20.1 percent (30.2 to 50.3) (Schrecker, 2010). Since September 11. 2001, academic freedom has been under attack. This approach has forced a hyper-professional "objective and neutral" perspective onto countless systemic issues of capitalism (i.e. race, gender, class and ability). Given the limited job security of most professors, speaking outside the accepted norm is becoming an increasingly risky endeavor.

Privatized Governance and Research to Increase Corporate Benefit

Lifelong academics and student-driven interests no longer govern universities despite the handful of professors who sincerely seek to inspire their students. Corporate interests dominate the makeup as well as the decisions made by boards of trustees (Giroux and Myrsiades, 2001). Most corporate giants have some form of specialized relationship with a research university through indirect connections and research firms. Others prefer to sit on the governing board of trustees directly and in doing so openly compromise the fundamental interests of the institutions. A significant proportion of grants comes from the private sector, resulting in the production of new knowledge and technologies to serve corporate interests over public interests.

The Spontaneous Development of Occupy Colleges

A multitude of factors within the neo-liberal education can be seen in part as the source of the current crisis of knowledge. "No Child Left Behind" standardized our generation on the basis of test scores while war and corporate tax breaks have bankrupted our collective future. Vast austerity measures in state funding have resulted in massive cuts in personnel, financial aid, and the diversity of programs. This was only further exacerbated by the 2007 financial collapse on Wall Street and corresponding "bailouts." With fewer dollars to support public colleges and universities, state administrators have willingly raised the cost of tuition, forcing students and their families to rely even more heavily on loans to finance higher education. The deregulation of the banking industry, in turn, further allowed banks and other institutions to exploit the dependency of students by demanding crippling interest rates. To make matters worse, the failing economy made it increasingly difficult for college graduates to find a job that pays a living wage, let alone make enough money to cover their student-loan debt.

All of this has created a perfect storm for the growing student movement in the United States, which has steadily gained momentum over the last three years. In 2009, students at the New School in New York seized the Graduate Facility the same semester students took over Campbell Hall at UCLA and planted the radical seeds of the Occupy movement. The 2010 U.S. Social Forum in Detroit provided youths an opportunity to solidify networks of resistance (e.g. New SDS) against the imperatives of corporate liberalism. Regional and state-based alliances have become increasingly common ever since. The network of dissenting youths had been steadily growing, and Occupy Wall Street (OWS) revealed the overwhelming size of the informal network.

In an attempt to spread the movement against corporate personhood and private interests to campus, students allied themselves with alumni and faculty to establish the Occupy Colleges initiative. The economic burdens created by tuition hikes across the country made it easy for students to adopt a parallel campaign. OWS came together to fight economic injustice, and students echoed this call, resolving to resist the injustice of corporatism on campus. Most U.S. colleges are modeled after the Wall Street profit model, with no representation from the actual stakeholders in the organization. It was our goal to bring the 99 percent mentality to the institutions of higher learning and aid in facilitating student resistance.

The Occupy Colleges initiative officially began on October 2, 2011 (just two weeks after OWS), as a Twitter handle and a Facebook group. After attending the new encampment in Los Angeles (Occupy LA), and watching the live broadcast of OWS general assembly's (GA), we recognized the need to bring the occupy movement to college students across the country. Campuses, like most cities, have town squares or plazas that would serve as our personalized Zuccotti Parks. For most students, their schools are big cities, community centers, or town squares that mirror the parks and meeting spaces occupiers were transforming across the country. Numerous encampments were established in conjunction with local Occupy College groups. However, in doing so, our talking points differed since we viewed the egregious tuition hikes and out-of-control student debt as the biggest economic injustices facing students. Occupy Colleges initiated its first direct action campaign by calling for a Nationwide Student Walkout on October 5, 2011, in which over 8,000 students demonstrated across the country.

The group of strangers quickly evolved into a strong interstate network of like-minded activists. As a collective whole, the student movement within the U.S. has been diverse in its establishment of specialized platforms unique to the conditions of each campus. However, the current issue uniting all students is the expanding cost of a diluted education. On a whole, the collective effort should ultimately be seen as a rejection of cynicism, apathy, electoralism, and the ennui of commercial culture.

The Principles of Occupy Colleges

From the beginning, the Occupy College's national GA agreed to stand in solidarity with the principles set forth by the OWS GA. As an affinity group of the national Occupy Movement, we felt it important to stay true to the beliefs that inspired the resurgence in student resistance. The national GA often referred to these principles in broad terms, and our network chose to emphasize the following points as topics of discussion.

> Engaging in direct and transparent participatory democracy; exercising personal and collective responsibility; recognizing individuals' inherent privilege and the influence it has on all interactions; empowering one another against all forms of oppression; redefining how labor is valued; the sanctity of individual privacy; the belief that education is a human right; and making technologies, knowledge, and culture open to all to freely access, create, modify and distribute. (Occupy Wall Street General Assembly Working Group, 2011/2012)

Solidarity

Like all networks of resistance, the global student movement is based on trust. As students (and former students) we continue to sow seeds of discontent, which in fact represent a broader interest (beyond that of private capital) to create a more egalitarian and sustainable society (Boren, 2001). From the unlimited general strike in Quebec, to the mass individual demonstrations throughout Korea, to the uprisings in Chile, and Yo Soy in Mexico, student movements have begun to constitute a considerable political force. We trust each other and seek to foster the success of one another by refusing to engage in negativity against members of our movement.

Autonomous Organization

Participating in the Occupy Colleges Movement was and remains entirely voluntary. The initiative lacked a single executive. Instead, it operated through a series of action committees. Each student chapter retains autonomy to challenge corporate or state power however they deem fit. Non-hierarchical relations between chapters granted every student the opportunity to participate. Much like OWS, the network connections and common identity were often our most valuable resources.

Consensus

The Occupy Colleges initiative brought with it the general assembly model of decision-making. However, student GAs have been described as less formal than city-based occupations. Much like OWS, a horizontal structure of organizing characterized the collective effort (Gelder and Positive Futures Network, 2011). As students mature into professional activists, it seems likely that the census model is likely to follow.

Direct Action Tactics

As a student movement, Occupy Colleges never felt a need to officially proclaim our support for direct action. While most people would consider our strategy to be nonviolent, a number of culture jamming tactics have stirred controversy (see Kalle Lasn's *Culture Jam*, 1999). Given the diversity of groups affiliated with the initiative, one would expect to find a range of perspectives on tactical issues. Some students were relatively moderate, seeking to inspire main-street reform through traditional avenues of electoral politics. Others preferred to revive the more militant approach of mass demonstrations and physical occupation of university property. These differences generated considerable disagreement regarding the organization's principles and its corresponding pledge of nonviolence. Organizers from the Center for the Working Poor in Los Angeles first adopted a pledge that was later offered by Occupy College to its members. However, the decision to formally endorse it was left to the individual campus chapters. Some eagerly agreed to the pledge, while other chapters refused. Within OWS, this was also a highly contested point in the Hedges vs. Black Bloc debates over destruction of property (see *Black Bloc, White Riot* by Thompson, 2010). To date 73 of 200 participating colleges have signed onto the nonviolence pledge.

Goals Adopted by Occupy Colleges

In the beginning of Occupy Colleges, it was important to adopt clearly defined goals we could all support while attempting to distinguish ourselves as the student wing of Occupy. Like numerous Occupy initiatives, we were not alone in our attempt to emancipate education. Occupy Education sought to promote critical education through Internet-based dialogue. Occupy Student Debt and Occupy Student Debt Campaign both sought to strictly target debt issues. After a few months of discussion, the ever-changing general assembly concluded that we were ultimately seeking to promote increased accessibility as well as a better quality education throughout the entirety of the U.S. collegiate system. Further, the national student assembly unanimously voted in favor of a macro-structural perspective, suggesting that we are only a small part of a larger global student movement with whom we stood in solidarity. Through their respective chapters, participants autonomously organized their campus and collectively campaigned to transform the educational paradigm as a whole. Many of the chapter representatives would never actually meet one another yet would correspond by orchestrating simultaneous protests with shared messages.

Education is a human right, a public good, and should be open to everyone. While Occupy Colleges facilitators believe in and actively seek free higher education for all, we envision a long road of struggle before this goal will be achieved. With one in five students defaulting on their loans, students feel as if an education is no longer "worth it" and are therefore avoiding it altogether (Perry, 2008). We demand that a tuition freeze be implemented throughout the country. From here we seek the elimination of all interest on student loans as well as re-implementation of debt protection permitting bankruptcy on both private and federal loans. Through a multitude of direct actions and mass demonstrations, we seek to eliminate all debt and institute a participatory model of education. Financial transparency into all university accounts, including the "special donation funds" often kept from students, will prove crucial in bringing about significant education reform. Until higher education is adequately subsidized (free of cost) by the state and the institutions are liberated from private interests of corporate control, the U.S. student resistance movement is likely to continue.

We are not consumers. We are students seeking a quality education. The business model of education exists at all policy levels (classroom, departments, administrators, board of trustees, board of governors, state and national legislatures). Student chapters have begun to organize their campuses against the corporate coup d'état in hopes of displacing careerist administrators whose interests compromise the foundational principles of public learning. In its place, we intend to install faculty-run universities on behalf of students, which will include meaningful student participation in campus governance.

As participants, students could serve an active role in budgeting, business practices, and historical development of their alma maters. We want funding to be redirected from the extracurricular nonsense (e.g., multi-million-dollar chancellor mansions), instead focusing on the enrichment of the collegiate expansion of knowledge. Faculty appointments need not only increase in number but in general quality in that there need to be more tenured faculty and more distinguished professorships. A successful movement would ultimately free up professors for outside-the-classroom student relations, class preparation, and hopefully research.

Social Media and Activism

Social media has served a monumental function in recruiting for our initiative. Occupy Colleges emerged and coordinated through these communication channels to learn and exchange ideas within our loose network of student activists. These tools were means to creating instances of spontaneous student power. However, the issue remains a complex topic of debate. Clicktivism is the pollution of activism with the logic of consumerism, marketing, computer science, and should not be seen as an end in itself (White, 2010). As a passive form of engagement, clicktivism has done little to engender social change yet has permitted the expansion of the movement.

> Social media sites such as Facebook and Twitter have been central organizing locations for spreading information about Occupy Wall Street . . . . Major uses of Facebook within the movement include the recruitment of people and resources to local occupations, information sharing, and storytelling and across-group exchanges. (Caren and Gaby, forthcoming)

Both sides in this debate have value since Occupy Colleges would never have gained national attention without this new form of activist communication. It would have been nearly impossible to inspire a mass walkout at 170 schools in three days if it had not been for obsessive Tweeting and Facebook conversations. However, too many students feel that simply utilizing social media was enough. From our experience, clicktivism is only useful when it leads to on-the-ground direct action.

Social media is not a catch-all since it fails to reach students at all class levels. The assumption that everyone has access to a computer or a smart phone is false. However, we have found that most students have access to cell phones, which permits mass communication to other students when protests have reached a point of contention with the administration or police. In fact, research has found that over 96 percent of young adults between 18 and 29 now own a mobile phone (PEW Internet and American Life Project, 2010). One of the best examples of this was the destruction of the UC–Davis camp. The moment officers started destroying tents on campus, students used the mass text-message system to alert reinforcements and in minutes hundreds of students showed up to defend their camp. Students were armed with cell phones equipped with cameras to witness the brutal pepper-spray attack perpetrated by Officer John Pike of the UC Davis Police Department. Within hours, videos and pictures of the attack went viral. Across the nation, students and concerned citizens began calling campus police to file formal complaints about the violent actions.

In addition to the use of cell phones, a new interactive GA model was offered through Inter-Occupy that allowed the Occupy Colleges network to transcend local confinement or geographical space. Starting in late December, students began conference calls to coordinate a nationwide demonstration in support of education. Students called for initiatives and the GA announced them throughout our communication channels. Some calls to action were popular while others went unheeded. "Maestro" conference systems permitted representative correspondence with the national general assembly (1 Stack, 2 Twinkles/Agree, 3 Disagree, 4 Direct response, 5 Technical support). The strongest participation on conference calls often came at times of student repression and leading up to national days of action.

Occupy Colleges Initiatives

Numerous campaigns quickly spread from campus to campus throughout the year and certain days of action were particularly successful. The initiative lacked momentum up until the national student walkouts and demonstrations called for on October 5th, 13th, and November 17th. The goal was to "Demand Education for the 99%." A number of student chapters began as a result of the acquaintances that came out of these three demonstrations. November 17th saw a new alliance with labor unions both on and off campus. In New York City alone, 30,000 students marched throughout the streets beneath the monolithic towers of power. Other events include mic-checking numerous bank CEOs, demonstrating against business departments, and organizing large-scale tuition stand-offs.

The next effort was to revive the teach-in popularized by Students for a Democratic Society (SDS) in the latter 1960s. On November 2nd and 3rd, 2011, 124 student chapters convened teach-ins in classrooms, public squares, and on state legislature lawns. Sharing in the educational tradition of the teach-in, professors were encouraged to connect with students, teachers, and the broader community in a non-hierarchical manner. However, most chapters reported that few professors remained dedicated to aiding the student movement. Technology permitted the live streaming (Ustream) of intellectual discourse on topics ranging from critical pedagogy to organizing against the Citizens United ruling. The live broadcasts of these public classes allowed viewers to participate all over the world. Through occupycolleges.org, viewers jumped from campus to campus listening in on these critical discussions. This radical approach to knowledge production was used again on February 22nd and 23rd, 2012.

The "March 1st Day of Education" was the first true inter-organization collaboration between education-related Occupy groups since Occupy's inception as a specialized movement. This laid the groundwork for Occupy Colleges' merger with Occupy Student Debt, which followed in the coming months. Seventy-three campuses held demonstrations focusing on austerity measures in state budgets. A number of our most successful demonstrations were made possible with the help of alumni or fellow off-campus occupiers. This was ultimately the last big national day of action before final exams.

Repression on Campus

Nightsticks, pepper spray, tasers, rubber bullets, and random punches have become normalized on U.S. campuses. Through our collective experiences, we have identified two distinct forms of repression experienced by students on campus. Violation of so-called free-speech zones brought threats to academic standing and physical encroachment from police officers. Through mass grassroots mobilizations, students fought back by staging protests, teach-ins, and rallies in support of democratizing their education. However, a number of these large-scale demonstrations resulted in students being seriously assaulted by the police.

"Free-Speech Zones" and Threats to Academic Standing

The form of repression often neglected is the marginalization of youth opinion. Students pay for universities in the form of taxes and tuition, yet are often contained to the peripheral areas of power. Particular areas are designated for small demonstrations while preventing the message from ever reaching the targets we are seeking to influence. These humiliating stipulations now go a step further and force students to have a permit to post fliers or demonstrate. Universities often view democracy as existing in particular areas; mandatory observation of these regulations must be heeded; otherwise permit holders will be aggressively pursued. The concept is in itself absurd. Students participating in mass demonstration on campus have not only seen threats of arrests or brutality but also threats to their academic standing (Nocella, Best, and McLaren, 2010). Students who demonstrated beyond their free-speech zones were often subjected to punishment from administrations. During the UC Board of Regents, students caught protesting were required to attend "free-speech classes" demanded by the Regent president himself, Mark Yudoff. Similar offenses against free speech occurred throughout California and the rest of the country.

Physical Encroachment

Military-grade pepper was one of the most widely used weapons against students. This "non-lethal method" was infamously exposed when a group of young females were kettled behind orange nets and pepper sprayed by Sergeant Bologna of the NYPD. This would be repeated during the UC–Davis incident weeks later. The resurgence in student activism also saw the return of on violent repression on campus. Students who stood up for their rights during OC direct actions faced multiple physical and legal obstacles. Documented cases of violence erupted from UC–Berkeley to Indiana University as students were confronted by police and administrators attempting to downplay the voice of students. Numerous other acts of brutality and repression occurred throughout the country; however a few particular conflicts managed to garner mass attention.

November 9: UC–Berkeley

The students of Occupy Cal at UC–Berkeley staged a walkout against rising tuition costs shortly after noon. The unarmed students surrounded Sproul Hall and were beaten with batons by university police officers in response. This was one of the first instances of violence in which university police openly squared off directly with students. These same acts of aggression that characterized the struggle for Zuccotti Park had reached the university in about a month's time.

November 18: UC–Davis Pepper Spray Incident and Walk of Shame

One of the most important displays of militant nonviolence came days after Officer Pike stepped over a line of seated students and pepper sprayed them at close range. UC–Davis Chancellor Katehi attended a closed-door meeting to discuss the attack the following day. Upon leaving the meeting, she was confronted by thousands of silent students who quietly watched as she walked to her car. Days after, Occupy Colleges organized a solidarity march on 69 campuses across the country in support of the brutalized students.

November 21: CUNY–Baruch College

Students protested a large tuition increase that had been levied across the entire CUNY (City University of New York) system. During the Board of Trustees meeting, students quickly gathered and occupied the building. Shortly thereafter, students were confronted by police officers wielding batons and seeking to make arrests. It quickly became apparent that student activists were being met with the same force they had experienced blocks away on Wall Street.

November 30: Kelly School of Business, Indiana University

Twenty students protested in solidarity with Occupy Wall Street when JP Morgan recruiters came to campus. Five students were shoved and pushed by campus police officers while students armed with cell phones captured the violence and posted it on YouTube within hours.

January 19: UC–Riverside Board of Regents

Projectile guns that shot rubber bullets and bean-bag ammunition were used against the students at UC–Riverside. Visual imagery of bruised arms and legs circulated on Twitter and other social media sites. A staff member at UC–Riverside recounts, "Someone moved toward the police—not aggressively, just like they were pushed—and the police started shoving students with batons, shot a few rounds, and arrested at least one student, whom 3 cops beat on the ground before they dragged him off" (Abrams, 2012).

February 10: UNC–Chapel Hill State Board of Governors Meeting
Another showdown occurred during a meeting in which the Board of Governors ordered a meeting in which all North Carolina public universities were ordered an 8.8 percent system-wide tuition hike. Students interrupted the meeting with a series of mic-checks while banging on the walls in hopes of disbanding the session. A student advocate against the hike was tackled and dragged into a nearby hallway where he was arrested and faced charges that were later dismissed. Students eventually gained entry to the meeting where they chanted from the floor and held a student general assembly.

Suggestions for Resisting Police Brutality on Campus

Through our experience as facilitators of Occupy Colleges, we discovered a few ways to combat police brutality on campus. Power in numbers, independent media coverage, avoiding infiltration, and the affinity group model for direct action, all help keep students safe when faced with police brutality. Large crowds that outnumber officers tend to prevent police violence. We suggest preparing mass text-message alerts as seen at the UC–Davis demonstrations. Preparation and spontaneous growth can quickly turn a handful of students into a full-blown demonstration with the energy to chase the administration off campus.

Likewise, if officers know they are being recorded they are likely to act in a less violent manner. It is crucial to obtain evidence of actions to be redirected against the violent administration. Our associations exist on an individual basis; however, infiltration of your student group can slow any campaign especially when using an open stack for discussion. In addition, infiltration runs the risk of leaking information to the campus police or opposition groups. Keep direct actions on a need-to-know basis between trusted individuals.

We highly suggest establishing affinity groups, similar to the buddy system, in which participants network and are held responsible for one anothers' security during direct actions. An affinity group is not a permanent structure but rather a convenient decision-making model that pools the interest of its participants. Groups can range anywhere from two to roughly 15 people; any larger and the groups runs into difficulties communicating. These census-based groups are autonomous and offer the opportunity for members to adjust and fragment as they see fit. Multiple affinity groups are often referred to as a "cluster." These groups serve as vehicles to execute an action while maintaining secrecy based on trust and clear communication. Affinity groups tend to pre-plan for a series of different scenarios and adjust accordingly through the action.

The Future of Occupy Colleges

The Occupy Colleges initiative has been in many ways successful during the emergence of a larger student movement (off-shoots such as Student Power, the expansion of New SDS and the renewed efforts of student syndicalism). Millennial youths in the U.S. are only at the beginning of their resurgence and have a long way to go to catch up to the students of Mexico, Chile, and Quebec. The goal now is to escalate the struggle and build broad alliances of student power. Coalitions across issue and state lines link independent struggles into a network with increased flexibility to adapt to new situations which need not necessarily fly an Occupy banner. Both location and issue orientation are grounds for networking. We advocate that students focus on creating overwhelming instances of student power aimed at dislodging corporate representatives within the administration. One such example is the recent popularization of divestment campaigns within the environmental and students for justice in Palestine efforts.

We believe that finding a balance between solidarity and autonomy will prove to be crucial in establishing a real movement. A U.S. student network is not only possible but necessary to the facilitation countless other causes beyond education. Such an organization could be incredibly loose and based at the state level with campus chapters serving as a part of the larger student union. Chapters are permitted autonomy to do as they please while collaborating within a grander movement of student resistance. A number of strong state-based networks currently have the unique opportunity to unite in the largest collective student struggle since 1968. To limit our potential aspirations as "too advantageous," neglects the historical significance of our opportunity.

> Every spring, more will join them from the next cohort of 400,000 unemployed graduates. Some will be embittered, and having nothing, will have nothing to lose. They will man the barricades and be the militants of a larger movement of increasingly impoverished working-class Americans. (Zimmerman, 2012)

The Wall Street speculators have and will continue to gamble with our futures just like they did with our childhood homes. We advocate that you seriously reconsider how you as an individual will seek to further your education, no matter your age, for we are all students. Those that have not been afforded the privilege of college deserve to occupy it the most.

Bibliography

Abrams, N. (2012, January 20). Students once again severely beaten by police. Occupy Colleges. Retrieved from http://occupycolleges.org/2012/01/20/press-release-students-once-again-severely-injured-by-police/

Boren, M. (2001). *Student resistance: A history of the unruly subject*. New York: Routledge Press.

Davidson, C. (1967). *Toward a student syndicalist movement: The new radicals in the multiversity and other SDS writings on student syndicalism*. Chicago, IL: Kerr Publishing.

Freire, P. (2005). *Teachers as cultural workers: Letters to those who dare to teach*. Cambridge, MA: Westview Press.

Gaby, S., and Caren, N. (forthcoming). Occupy online: How cute old men and Malcolm X recruited 400,000 U.S. users to OWS on Facebook. *Social Movement Studies*.

Gelder, S. V., and the Positive Futures Network. (2011). *This changes everything. Occupy Wall Street and the 99% Movement*. San Francisco, CA: Berrett-Koehler Publishers.

Giroux, H. A., and Myrsiades, K. (2001). *Beyond the corporate university: Pedagogy, culture, and literary studies in the new millennium*. New York: Rowman & Littlefield.

Greenhouse, S. (2012, May 12). Jobs few, grads flock to unpaid internships [editorial]. *New York Times*. http://www.nytimes.com/2012/05/06/business/unpaid-internships-dont-always-deliver.html?pagewanted=all

Harris, M. (2011, May 9). Rising tuition + student loans = education bubble. *The Fiscal Times*. http://www.thefiscaltimes.com/Articles/2011/05/09/rising-tuition-student-loans-education-bubble.aspx#page1_-

Kaplan, D. (2012, May 10). The draconian hidden penalty on student loans. *CNN Money*. http://finance.fortune.cnn.com/2012/05/10/student-loan-penalty

National Center for Education Statistics at the Institute for Education Science. (2012). Projection of education statistics to 2020. http://nces.ed.gov/programs/projections/projections2020/tables/table_33.asp

New York City General Assembly. (2011). Principles of solidarity. http://www.nycga.net/resources/declaration/

Nocella II, A. J., Best, S., and McLaren, P. (2010). *Academic repression: Reflections from the Academic-Industrial Complex*. Oakland CA: AK Press.

Occupy Colleges. (2012). Student once again severely injured by the police [Press release]. http://occupycolleges.org/2012/01/20/press-release-students-once-again-severely-injured-by-police/

Occupy Wall Street General Assembly Working Group. (2011/2012). Principles of solidarity. http://www.nycga.net/resources/principles-of-solidarity/

Perry, N. (2008, October 6). With no way out of trouble, more students likely to default. http://seattletimes.nwsource.com/html/localnews/2008231488_loandaytwo06m.html

PEW Internet and American Life Project. (2010, October 18). Survey: 96% of young adults own cell phones. Retrieved from http://pewinternet.org/Media-mentions/2010/Survey-96-of-Young-Adults-Own-Cell-Phones.aspx

Schrecker, E. (2010). *The lost soul of higher education: Corporatization, the assault on academic freedom, and the end of the American University*. New York: The New Press.

Thompson, A.K. (2010). *Black bloc, White riot: Anti-globalization and the genealogy of dissent*. Oakland, CA: AK Press.

White, M. (2010). What is clicktivism? http://www.clicktivism.org/

Yen, H. (2012, May 23). 1 in 2 new graduates are jobless or unemployed. *Yahoo Finance*. http://finance.yahoo.com/news/1-2-graduates-jobless-underemployed-140300522.html

Zimmerman, B. (2012). The aftermath of Occupy will surpass the gains of 1960s activism. *Truth-out*. http://truth-out.org/opinion/item/10125-the-aftermath-of-occupy-will-surpass-the-sixties

• CHAPTER FOURTEEN •

Faculty Should Join with Occupy Movement Protesters on College Campuses

Henry Giroux

In the United States as well as many other countries, students are protesting against rising tuition fees, the increasing financial burdens they are forced to assume, and the primacy of market models in shaping higher education while emphasizing private benefits to individuals and the economy. Many students view these policies and for-profit industries as part of an assault on not just the public character of the university but also an attack on civic society and their future. For many young people in the Occupy Movement, higher education has defaulted on its promise to provide them with both a quality education and the prospects of a dignified future. They resent the growing instrumentalization and accompanying hostility to critical and oppositional ideas within the university. They have watched over the years as the university is losing ground as a place to think, dissent, and develop a culture of questioning, dialogue, and civic enlightenment. They are rethinking what should be the role of the university in a world caught in a nightmarish blend of war, massive economic inequities, and ecological destruction. What role should the university play at a time when politics is being emptied out of any connection to civic literacy, informed judgment, and critical dialogue, further deepening a culture of illiteracy, cruelty, hyper-masculinity, and disposability.

Young people are not only engaging in a great refusal, they are also arguing for the social benefits and public value of higher education, while deeply resenting the fact that as conservative politicians defund higher education and cut public spending, they do so in order to be able to support tax breaks for corporations and the rich and ample funds for sustaining and expanding the warfare state. The Occupy protesters view the assault on the programs that emerged out of the New Deal and the Great Society as being undermined as society increasingly returns to a Second Gilded Age in which youths have to bear the burden of an attack on the welfare state, social provisions, and a huge wealth and income inequality gap. Young people recognize that they have become disposable, and higher education, which always embodied the ideal, though in damaging terms, of a better life, has now become annexed to the military-academic-industrial complex. What is important about the Occupy protesters' criticism of being straddled with onerous debt, viewed as a suspect generation, subjected to the demands of an audit culture that confuses training with critical education, and their growing exclusion from higher education is that such concerns situate the attack on higher education as part of a broader criticism against the withering away of the public realm, public values, and any viable notion of the public good. To paraphrase William Greider, they have come to recognize in collective fashion that higher education has increasingly come to resemble "an ecological dead zone" where social relevance and engaged scholarship perishes in a polluted commercial market-driven environment. The notion of the university as a center of critique and a vital democratic public sphere that cultivates the knowledge, skills, and values necessary for the production of a democratic polity is giving way to a view of the university as a marketing machine essential to the production of identities in which the only obligation of citizenship is consuming.

The Occupy Wall Street protesters reject the propaganda they have been relentlessly fed by a market-driven culture: the notion that markets should take priority over governments, that market values are the best means for ordering society and satisfying human needs, that material interests are more important than social needs, and that self-interest is the driving force of freedom and the organizing principle of society. The Occupy Wall Street protests refuse a notion of society that embraces a definition of agency in which people are viewed only as commodities, bound together in a Darwinian nightmare that celebrates the logic of greed, unchecked individualism, and a disdain for democratic values. The old idea of democracy in which the few govern the many through the power of capital and ritualized elections is being replaced with a new understanding of democracy and politics in which power and resources are shared and economic justice and democratic values work in the interest of social responsibility and the common good. This radical notion of democracy is in the making, unfinished, and open to connecting people, power, resources, and knowledge. And this turn towards a radical understanding of connecting the particular to the general is particularly true of their view of higher education. What the Occupy protesters recognize, as the British educator Simon Dawes points out, is that "'the public university' can be read as shorthand for 'not-neoliberal university', where neoliberal means more than private funding; it means 'not good for democracy'" (2011, para. 20).

All over the country, Occupy Movement protestors are setting up camps on college campuses. Not only are they protesting the ways in which universities now resemble corporations treating faculty as a subaltern class of casualized labor and defining students largely as customers and clients. They have also recognized that banks and loan corporations with their army of lobbyists have declared war on students, killing any legislation that would reduce the cost of schooling, stifling any legislation that would make it affordable for all working- and middle-class students. They are also raising serious questions about academics. Where are they when it comes to protesting the corporatization and militarization of higher education? Why are so many of them complicit with the ideologies and money now used by corporations and the national security state to promote the interest of finance capital and agencies such as the CIA, Defense Department, Pentagon, and other apparatuses of the National Security State intent on recruiting students to produce militarized knowledge and create new and ever more sophisticated surveillance systems and weapons of mass destruction? Why do so many academics cling to a notion of disinterested and objective scholarship and publish and make a claim to pedagogy that allegedly decries any relationship to politics, power, or interest in larger social issues? What Occupy Movement protesters have recognized is that for all intents and purposes, too many academics who make a claim to objectivity and in some cases reject the presence of the military-industrial-academic complex on campus have become irrelevant to offering any viable defense of the university as a democratic public sphere; or for that matter, some even defend to a broader public the very conditions that make their work possible.

One important question that arises from the Occupy Movement's migration to college campuses is what can academics learn from these young people? One of the things they might learn is that critical and important forms of education and dialogue are taking place outside of the university in which issues are being talked about that are often ignored within the halls, disciplines, and classrooms in many universities. Many universities have lost touch with bridging the production of knowledge, research, and teaching with the myriad urgent social issues now facing the larger society, including crushing poverty, environmental degradation, racism, the suspension of civil liberties, the colonization of the media by corporations, the rise of the punishing state, religious fanaticism, the corruption of politics by big money, and more. Since the 1980s, higher education has been increasingly corporatized and militarized and subject to market-driven values and managerial relations that treat faculty and students as entrepreneurs and clients, while reducing knowledge to the dictates of an audit culture, and pedagogy to a destructive and reductive instrumental rationality. It is hoped that academic might both learn and be inspired about the current attempt on the part of students to change the conversation about the meaning and purpose of higher education. They might be moved and educated by the attempt on the part of many young people today to reclaim higher education as a democratic public sphere, one that not only provides work skills, but also offers a formative culture that prepares students to be critical and active agents in shaping the myriad economic, social, and political forces that govern their lives. Students are rejecting a model of education based on narrow forms of measurable utility, capital accumulation, and cost-efficient asset and power-stripping measures; they are rejecting a market-driven model of education that reduces 70 percent of faculty to a subaltern class of part-time workers, and treats students as customers and commodities, offering them overcrowded classrooms, skyrocketing tuition rates, and modes of learning that have little to do with enabling them to translate personal troubles into social problems. Universities increasingly have come to resemble malls and rather than offer students an education in which they can become critical individuals and social agents who believe that they have the power to change things, they are largely reduced to passive consumers entertained by the spectacles of big sports, celebrity culture, and the lure of utterly privatized desires.

Putting their bodies on the line, many Occupy protesters on college campuses, such as the Berkeley and Davis campuses in the University of California system, have been subject to a stream of police brutality, much of which has been caught on video and distributed over the Internet. For example, students at UC–Davis, while sitting down and protesting peacefully, were brutally attacked by police in riot gear who proceeded to cover them with military-grade pepper spray. One professor who witnessed the event claimed that the police actually knelt on some students and indiscriminately put pepper spray down their throats (Gibson, 2011, para. 4). Using their First Amendment rights to speak out, UC–Berkeley students were violently pummelled with batons and bean-bag guns (Friedersdorf, 2011). In Oakland, protesters were repeatedly attacked and beaten. In one particularly brutal incident, Iraq war veteran Scott Olsen was struck by what appeared to be a teargas canister in one melee and as a result suffered a traumatic head injury. Legitimate dissent and protests, acts that should be nourished on any university campus that takes real learning seriously, are now subject to police brutality, repression, and violence. But there is more at work here than the grotesque display of police brutality against students and peaceful protesters; there is also a striking snapshot of the increasing militarization of police forces in the United States and the growing support among the American public for criminalizing dissent and the use of intimidation, military-grade technologies, and overt force, as a "favored method for dealing with dissent of any kind" (Taibbi, 2011, para. 5). Glenn Greenwald

(2011) views such attacks as evidence of a police state in its purest form. He writes:

> The U.S. Government—in the name of Terrorism—has aggressively para-militarized the nation's domestic police forces by lavishing them with countless military-style weapons and other war-like technologies, training them in war-zone military tactics, and generally imposing a war mentality on them. Arming domestic police forces with para-military weaponry will ensure their systematic use even in the absence of a Terrorist attack on U.S. soil . . . . It's a very small step to go from supporting the abuse of defenseless detainees (including one's fellow citizens) to supporting the pepper-spraying and tasering of non-violent political protesters. (para. 6)

Such violence presents a direct threat to academic freedom, dissent, and the most elemental of civil rights. What the Occupy protesters have made clear through an act of collective defiance, resistance, and courage is that a state that is willing to torture foreigners will not hesitate to violate the rule of law in squelching dissent at home. And it is precisely this ongoing militarization and erosion of civil liberties and democracy itself that many of the Occupy protesters are resisting and collectively fighting with their courage, minds, and bodies. As this drama of increasing globalized protest and repression is played out daily, faculty should do everything they can to both protect students and to oppose the encroaching of police-state tactics as they encroach upon our most sacred civil liberties. "Academics as well as the general public must protest against the growing violence and militarization of everyday life that has become the most important element of power and mediating force in shaping social relationship in the United States. What faculty can learn from young people is the need for a politics in which the pain of others, particularly those considered disposable or unworthy, is the subject of compassion rather than ridicule. They can emulate young people who are protesting a culture in which death and suffering are now considered another form of entertainment designed to raise the collective pleasure quotient. They can join in solidarity with Occupy protesters who are rejecting a culture that not only valorizes pain and human suffering but turns it into a source of entertainment and pleasure, even when it has to do with the suffering of real human beings. Judith Butler is on target in her insistence that what young people are making clear all across the globe is that youthful protesters have been abandoned "[a]nd no democratic system can abandon its people when it claims to represent all its people" (Bella, 2011, para. 18).

In many ways, students are offering faculty the possibility of becoming part of a larger conversation, if not social movement, one that addresses what the role of the university might be in relation to public life in the 21st century. Central to such an inquiry is examining how higher education has been caught in the grip of larger economic and political forces that undermine the social state, social provisions, and democracy itself. The Occupy protestors are arguing that while they might support a limited notion of a market economy, they do not want to live in a market society, a society in which market values become a template for organizing all aspects of social life. They have learned the hard way that beneath this market fundamentalism resides a mode of education and set of values that contain a secret order of politics that is destructive of democratic social relations, democratic modes of equality, and civic education itself. Young people can make clear to faculty that over the last 30 years they have been written out of the social contract and are no longer viewed as a symbol of hope, just as they have been written out of the power relations that govern the university. No longer regarded as an important social investment or as a marker for the state of democracy and the moral life of the nation, young people have become the objects of a more direct and damaging assault waged on them on a number of political economic and cultural fronts. They have been deprived of decent scholarships, disrespected in their attempts to gain a quality education, foiled in their attempts to secure a decent job, and denied a voice in the shaping of the institutions that bear down heavily on their daily lives. Big banks and large financial institutions view them as a drain on the nation's financial coffers and as a liability in making quick financial profits through short-term investments. Young people are now challenging this toxic form of casino capitalism and in doing so are changing the national conversation that has focused on deficit reduction and taxing the poor to important issues that range from poverty and joblessness to corporate corruption. Put differently, the Occupy protesters are asking big questions and they are not simply being moralistic in that they are also demanding an alternative vision and set of policies to drive American society. Faculty need to listen to young people in order to try to understand the problems they face and how as academics they might be unknowingly complicit in reproducing such problems. They also need to begin a conversation with young people and among other faculty about how they can become a force for democratic change.

Young people need a space on campuses to talk back, talk to one another, and engage in respectful dialogue with faculty and learn how to engage in coalition building. Faculty and administrators can begin to open up the possibility for such spaces by offering the Occupy protestors an opportunity to speak to their classes, create autonomous spaces within the university where they might meet and engage in dialogue with others, and go even further by joining them in fighting those economic and political forces that are destroying higher education as a social good and as a citadel of rigorous intellectual engagement and civic debate. Young people no longer recognize themselves in terms preferred by the market and they no longer believe in an education that ignores critical thinking, dialogue, and those values that engage matters of social responsibility and civic engagement. But students have more to offer than a serious critique of the university and its complicity with a number of anti-democratic forces now shaping the larger society; they are also modeling for faculty new modes of participatory democracy, and exhibiting forms of pedagogy and education that connect learning with social change and knowledge with more democratic modes of self-development and social empowerment. Clearly, academics have a lot to learn from both the ways in which students are changing the conversation about education, important social issues, democracy, and what it might mean to imagine a new understanding of politics and a different future. All of these issues are especially true for those faculty members who believe that scholarship should be disinterested and removed from addressing important social issues. The questions students are raising are important for faculty to rethink those modes of professionalism, specialism, and social relations that have cut them off from addressing important social issues and the larger society. Professionalism does not have to translate into a flight from moral and intellectual responsibility. Faculty can also put pressure on their unions to support the Occupy Movement, provide them with financial and media resources, and join with them in pushing for educational and political reforms. The Occupy protestors are surely right in arguing that higher education is a vital public sphere that should be at the forefront in addressing important political, economic, and social issues; and faculty should combine their scholarly rigor and knowledge to bridge the gap between the university and everyday life, not to benefit corporate interests or the warfare state, but to benefit existing and future generations of young people who hold the key to whether democracy will survive the current moment in American history. Too many academics for too long have turned their back on addressing important social issues, joining with young people to fight with them for a better future, and using their knowledge and skills to convince a wider public that higher education is crucial not just for students, but that its benefits embrace the common good and the entire society. Joining with students in the Occupy Movement is not merely a career choice, it is a choice about what kind of society we all want to live in, and how the urgency of that question at the current historical moment demands that academics take that question seriously and act as quickly as possible with passion and conviction.

Students in the United States and in many other parts of the globe have become stateless, without rights, jobless, and without hope. In opposition to the conditions responsible for their being abandoned by governments that destroy democracies and reduce the state to an instrument of repression, it is crucial for academics to join with students to make possible once again the conditions for community, solidarity, and politics. Youths have become the space where repressive regimes deposit their nightmares, hoping to keep them contained within bubbles of consumption and celebrity culture, or worse in prison cells. Those days are over as youths become the new subaltern class capable of forging themselves with others into a new counter power and a force for real effective change. This is not a movement that faculty should ignore; and if they do so, it will be at their own expense because they might just be the next target that the punishing state decides to attack.

Bibliography

Bella, K. (2011, December 15). Bodies in alliance: Gender theorist Judith Butler on the Occupy and SlutWalk Movements. *TruthOut*. http://www.truth-out.org/bodies-alliance-gender-theorist-judith-butler-occupy-and-slutwalk-movements/1323880210

Dawes, S. (2011, November 22). The "public university" as a response to the funding cuts to higher education in the UK. *Truthout*. http://www.truth-out.org/public-university-response-funding-cuts-uks-higher-education/1321984014

Friedersdorf, C. (2011, November 10). UC Berkeley riot police use batons to clear students from Sproul Plaza. *The Atlantic*. http://www.theatlantic.com/national /print/2011/11/uc-berkeley-riot-police-use-batons-to-clear-students-from-sproul-plaza/ 248228/

Gibson, C. (2011, November 16). America has become a fascist police state. *RSN*. http://readersupportednews.org/opinion2/275-42/8599-carl-gibson-america-has-become-a-fascist-police-state

Greenwald, G. (2011, November 20). The roots of the UC-Davis pepper-spraying. *Salon*. http://www.salon.com/2011/11/20/the_roots_of_the_uc_davis_pepper_spraying/

Taibbi, M. (2011, November 22). UC Davis pepper-spray incident reveals weakness up top. *Rolling Stone*. http://www.rollingstone.com/politics/blogs/taibblog/uc-davis-pepper-spray-incident-reveals-weakness-up-top-20111122

Contributors

Natalia Abrams is an educational activist working toward ways to provide accessible, quality, affordable education for all students. Natalia's first engagement with activism came when demonstrating against the War in Iraq and Afghanistan. She continued her activism by majoring in women's studies at UCLA, where she was vice president and student intern of the Feminist Majority Foundation. Upon graduating UCLA, summa cum laude in 2009 education and student debt became the salient issues of Natalia's activism. After attending OccupyLA and seeing the beauty of the participatory democracy, Natalia co-founded the Occupy Colleges initiative, a national student movement. While serving as a media representative of Occupy Colleges, Natalia has written for the Student Nation and commentated for *The Dylan Ratigan Show* on MSNBC and CBS *Sunday Morning*. She has also been quoted in the *New York Times*, Huffington Post, *The Nation*, NPR, Reuters, AP, among many other media sources. Natalia is passionate about engaging with college students and tackling the issues of student debt.

Sarah Augusto is a Ph.D. Candidate in Sociology at the University of California, Davis. Her research and teaching interests include social movements, emotions, and sex and gender. She is currently writing her dissertation on the organization and internal dynamics of the UC student movement and the Occupy Movement. She has been active in the student movement at UC Davis since 2009.

Ben Brucato is an instructor at Northern Arizona University and currently completing a Ph.D. in Science and Technology Studies (STS) at Rensselaer Polytechnic Institute in Troy, New York. He turns his interest in politics of technology toward the ideological inscription of objects and the ecologies in which they are involved. An aim of his scholarship is to develop analytic and evaluative tools that will allow effective intervention into technical relations and the shaping of those that are more conducive to free communities. Recent publications have discussed surveillance at colleges and universities and the Occupy Wall Street movement in the context of the financial crisis. His current research investigates the intersection of ubiquitous surveillance and police violence.

Jason Del Gandio, Ph.D., (rhetoric4radicals@gmail.com) is an Assistant Professor of Public Communication at Temple University. He specializes in rhetoric, the philosophy of communication and language, critical analysis, radical social and political theory, and to a lesser extent, performance studies. He participated in the global justice movement in the early 2000s and went on to fair trade activism, Latin American solidarity, and anti-war activism. These experiences led Jason to write *Rhetoric for Radicals: A Handbook for 21st Century Activists* (November 2008), which received the Gold Award by the Independent Publishers Industry in the social activism category. He also offers workshops for activists on communication and rhetoric. He is now working on his next book, which focuses on the political relationship between desire and reality. Jason is a writer, thinker, activist, and teacher dedicated to social justice. More information can be found at www.jasondelgandio.com.

Scott DeMuth has been involved in several projects in the Twin Cities including the Anarchist Black Cross and the Jack Pine Community Center. He is currently a member of EWOK! (Earth Warriors Are OK!), Oyate Nipi Kte, and the editorial collective for the Dakota community journal *Anpao Duta*. He is a Dakota language student and a graduate student in the Sociology Department at the University of Minnesota.

Luis A. Fernandez is an Associate Professor in the Department of Criminology and Criminal Justice at Northern Arizona University. Dr. Fernandez received his Ph.D. from the School of Justice Studies and Social Inquiry at Arizona State University. He has worked for several research institutions, including the National Council on Crime and Delinquency, and is a founding member of the Institute for the Study of Dissent and Social Control (http://www.dissensio.org). His research and teaching interests include protest policing, social movements, globalization, and issues in the social control of late modernity. He is the author of *Policing Dissent: Social Control and the Anti-globalization Movement* (Rutgers University Press, 2008) and articles in various journals, including *Social Justice and Qualitative Sociology*. His most recent research examines the securitization of the Mexican–Guatemalan border as it links to US immigration policy. Among the courses he teaches are Global Justice, Law Enforcement Systems, and Research Methods.

David Gabbard, Ph.D., has earned national and international recognition for his work in critical educational policy studies and democratic educational theory. Along with five published books, his record of scholarly production includes over fifty articles and book chapters. The first edition of his *Knowledge and Power in Global Economy: Politics and the Rhetoric of School Reform* received the Critic's Choice Award from the American Educational Studies Association in 2001. He is currently working with more than fifty distinguished contributing authors in producing a second edition of that work, taking a new focus on 'The Effects of School Reform in a Neoliberal / Neoconservative Age.' Professor Gabbard has also co-edited *Education as Enforcement: The Militarization and Corporatization of Schools* (with Ken Saltman), *Defending Public Schools: Education Under the Security State* (with E. Wayne Ross) and *Michel Foucault and Power Today (with* Alain Beaulieu). Professor Gabbard also co-founded and co-edits *Public Resistance: An Academic Journal to Confront the Lies of the Right* with Karen Anijar-Appleton. He currently serves as Program Coordinator for the Marxian Analysis of Society, Schools, and Education Special Interest Group of the American Educational Research Association. He also sits on the editorial boards for *The International Journal of Inclusive Democracy* and *The Journal of Critical Education Policy Studies.*

Henry A. Giroux, Ph.D., currently holds the Global TV Network Chair Professorship at McMaster University in the English and Cultural Studies Department. His most recent books include: *Zombie Politics and Culture in the Age of Casino Capitalism* (Peter Lang, 2011), *On Critical Pedagogy* (Continuum, 2011), *Twilight of the Social: Resurgent Publics in the Age of Disposability* (Paradigm, 2012), and *Disposable Youth: Racialized Memories and the Culture of Cruelty* (Routledge, 2012). His most recent book is *Youth in Revolt: Reclaiming a Democratic Future* (Paradigm, 2013). His website is www.henryagiroux.com.

Caroline K. Kaltefleiter, Ph.D., is Coordinator of Women's Studies and Professor of Communication Studies at the State University of New York College at Cortland and a Mary Lily Research Fellow at Duke University. She has over twenty years of media experience as a news anchor and producer for public and community radio stations. She produces media programs and research related to youth culture and popular culture. She is the anchor and executive producer of *The Digital Divide,* a radio program devoted to new technology and activism. An activist and an academic, Kaltefleiter is the author of "Revolution Girl Style Now: Riot Grrrl Practices and Action" in *Anarchist Studies* (Routledge, 2009). Her forthcoming book on the Riot Grrrl movement examines the girl movement and its use of alternative media forums such as zines, street demonstrations, internet mash-ups, and Do-It Yourself (DY) musical recordings. She can be reached at caroline.kaltefleiter@cortland.edu.

Joe Lewis began his career in 1991 as a high school English teacher in Wisconsin. Since then, he has taught secondary and university- level English and English Education in a variety of places and positions, including two years in Morocco and four years in New York City. Joe holds a doctorate in English Education from Teachers College, Columbia University. His areas of scholarly interest and writing include: the teaching of reading and writing; cross-cultural language and literacy practices; narrative ethnography; post-modern, postcolonial, and de-colonial forms of ethnography. Joe is an Assistant Professor of English Education at Hamline University in St. Paul, MN.

Barbara Madeloni is a senior lecturer in the School of Education, the University of Massachusetts Amherst and a member of the Education Radio collective. After receiving her doctoral degree in Psychology at the University of Denver and practicing as a psychotherapist for 15 years, Barbara became increasingly alarmed by the erosion of democratic potential in the US. She returned to school, earned her Masters in Education, and became an English teacher with the goal of strengthening and protecting the possibilities for democracy. Her commitments are to teaching, scholarship, and activism that embody education as liberation. Along with the letter of non-renewal discussed in her chapter in this book, Barbara was admonished to always be clear that she does not speak for the University. She would like to make it clear that she does not speak for the School of Education of the University of Massachusetts Amherst.

Anthony J. Nocella II, scholar–activist, is a Visiting Professor in the School of Education at Hamline University and the Executive Director of the Institute for Critical Animal Studies and Chairperson of Save the Kids. He received his doctorate from the Maxwell School at Syracuse University and is currently the editor of *the Peace Studies Journal* and *Transformative Justice Journal*. He has been involved in numerous political campaigns, organizations, and international demonstrations fostering direct democracy and is a co-founder of more than ten active political organizations and five scholarly journals. He has contributed to more than two dozen publications and has published more than eleven books, including *A Peacemaker's Guide for Building Peace with a Revolutionary Group* (PARC, 2004). He has co-edited (with Steve Best) *Terrorists or Freedom Fighters? Reflections on the Liberation of Animals* (Lantern Books, 2004) and *Igniting a Revolution: Voices in Defense of the Earth* (AK Press, 2006). More recently he has edited *Contemporary Anarchist Studies* (Routledge, 2009). Nocella can be reached through his website: www.anthonynocella.org.

Christian Parenti is a correspondent for *The Nation* and is author of *The Freedom: Shadows and Hallucinations in Occupied Iraq* (New Press, 2004). He received a Ph.D. in sociology from the London School of Economics in 2000. His two previous books are *The Soft Cage: Surveillance in America from Slavery to the War on Terror* (Basic Books, 2003) and *Lockdown America: Police and Prisons in the Age of Crisis* (Verso, 2000). He has been a Soros Senior Justice fellow and a Ford Foundation Fellow at the CUNY Graduate School's Center for Place, Culture, and Politics.

David Naguib Pellow, Ph.D., is Professor of Sociology at the University of Minnesota. His teaching and research focus on environmental justice issues in communities of color in the U.S. and globally. His books include: *The Slums of Aspen: Immigrants v. the Environment in America's Eden* (with Lisa Sun-Hee Park), *Resisting Global Toxics: Transnational Movements for Environmental Justice*; *The Silicon Valley of Dreams: Environmental Injustice, Immigrant Workers, and the High-Tech Global Economy* (with Lisa Sun-Hee Park); and *Garbage Wars: The Struggle for Environmental Justice in Chicago*. He has served on the boards of directors for the Center for Urban Transformation, Greenpeace USA, and International Rivers. He is currently the facilitator of the Minnesota Global Justice Project.

Julie Setele is a Ph.D. Candidate in Sociology at the University of California, Davis. Her research and teaching interests include critical criminology, social welfare, and reproductive politics. She is fascinated by the reproduction of inequality, especially within the criminal justice, welfare, and educational systems. Her dissertation explores criminal justice policy-making at the local level, focusing on the roles of formerly incarcerated people in shaping reentry policy. As a Civility Project fellow at the UC Davis Humanities Institute, she recently completed a study of institutional and community responses to the wave of hate and bias-related incidents that occurred on UC campuses in 2010. A web-based presentation of the findings is available at http://civilityproject.ucdavis.edu/. With experience teaching college students at both UC Davis and San Quentin State Prison, Julie has served as a teaching consultant for the UC Davis Sociology Department and the UC Davis Center for Excellence in Teaching and Learning. She would love to have a full-time teaching job in a truly free and democratic public university.

Kim Socha grew up in Philadelphia is now a professor in the Twin Cities, Minnesota. Holding a Ph.D. in Literature, she is an English instructor at Normandale Community College with scholarship on topics such as critical pedagogy, surrealism, and Latino/a literature. Her activist experience includes assisting survivors of domestic violence and sexual abuse, drug abuse, and addiction in their recoveries. She is also part of a transformative justice program at the Minnesota Department of Corrections. Her 2012 book *Women, Destruction, and the Avant-Garde: A Paradigm for Animal Liberation*—an intersectional study investigating feminism, race, animal liberation and the arts—was published through the Institute for Critical Animal Studies' Rodopi book series. Kim has co-edited and contributed to a book by activists local to the Twin Cities (McFarland, 2012).

Maura Stephens is a long-time journalist, theater artist, playwright, educator, and peace, justice, and environmental activist. She learned the craft of journalism during 19 years in various editorial capacities at *Newsweek* and *Newsweek International* magazines, leaving to co-found an early Internet startup and move into independent media. From 1997 to 2010 she was publishing editor of *ICView,* the magazine of Ithaca College. She is currently associate director of the Park Center for Independent Media at Ithaca College and teaches professional and citizen journalism. She writes on a wide array of subjects including Iraq, Burma, international relations, social justice, politics, and environment for such independent media as Alternet, Common Dreams, TruthOut, and *Yes!* She founded or co-founded five antifracking groups and is finishing the last rewrite of her book *Frack Attack: Fighting Back,* about the dangerous and destructive industrial process that threatens 27 U.S. states and many countries around the globe.

Wes Strong is an activist and a writer based in Hartford, Connecticut. Wes has been highly involved on the national level with the movement to defend public education for two years. He has been active in anti-war, workers rights, and anti-racist struggles since 2006. Wes graduated with honors from Central Connecticut State University (CCSU) in 2008 where he was a member of a peace group that was being monitored by the federal government along with hundreds of other similar organizations. He participated in several actions at CCSU in opposition to the university newspaper's printing of racist and sexist material, which attracted national attention in 2007. Wes played a role in building major actions on March 4th, 2010, and continues this work to this day, developing national-level coordination. Wes is a regular contributor and editor of New Student Left Review which covers politics and social movements from a leftist perspective http://nslreview.net.

Ryan Thomson is a twenty-two-year-old scholar-activist currently working towards a Ph.D. in Sociology at North Carolina State University in Raleigh. His areas of research specialization include global and rural issues with an emphasis on public engagement. Ryan's current extension research with the NC EPA emphasizes spatial inequality and disproportionate patterns of pollution. He currently sits on the board of NC Waste Awareness and Reduction Network and works closely with Democracy North Carolina. Prior to his experiences in Zuccotti Park and co-founding the Occupy Colleges initiative, Ryan was a student organizer for a diverse range of issues promoting social and ecological justice. During a series of apprenticeships and as campus coordinator for the Greenpeace Student Network, he co-founded the NCSU Students for a Democratic Society (SDS) and NCSU Environmental Concerns Organization (ECO). In tandem with the larger Defend Education and Powershift movements, these two direct-action organizations have been highly active in building student opposition to the corporatization of higher education and the industrial degradation of regional ecosystems.

Richard Van Heertum is currently a lecturer at the Art Institute of California in North Hollywood and Drexel University, having previously worked as a visiting professor at UCLA and CUNY. He received his Ph.D. in education and cultural studies from UCLA in 2009 and completed a Masters in economics at San Diego State in 1997. He has published over 25 academic essays, over 130 articles in the popular press and co-edited two books, *Hollywood's Exploited* (Palgrave, 2010) and *Educating the Global Citizen* (Bentham, 2011) as well as presenting at numerous conferences. He previously served as the Program Officer for the Paulo Freire Institute at UCLA and is a member of AERA, NAMLA, PES, ACME, CAPE, and the National Arts Club. His work focuses on urban educational policy and reform, the political economy of globalization, social justice education and the politics of popular culture.

Daniel White Hodge, Ph.D., is the Director of the Youth & Adolescent Culture Center and Assistant Professor of Youth Culture at North Park University in Chicago. He has taught at Cal State University Northridge's religious studies department as well as Fuller Theological Seminary's school of intercultural studies. As a speaker, writer, and activist, he has spoken on many college campuses including Stanford University, UCLA, USC, and Union Theological Seminary. Dan teaches classes around the world on subjects such as Black popular culture, personality & the self, Hip Hop discourse, and race/ethnicity within religion. His two current books are *Heaven Has a Ghetto: The Missiological Gospel & Theology of Tupac Amaru Shakur* (VDM, 2009), and *The Soul of Hip Hop: Rimbs, Timbs, & A Cultural Theology* (IVP, 2010). He is currently working on a book titled *The Hostile Gospel: Finding Religion in the Post Soul Theology of Hip Hop* (Brill Academic, 2012). He has published in *Youth Worker Journal*, *Youth & Family Journal* and has several book chapters in various readers. Dan and his wife Emily are most proud of their 5-year-old dancing daughter, Mahalia Joy.

Index

• E •

• F •

• G •

• H •

• I •

• J •

• K •

• L •

• M •

• N •

• O •

• P •

• Q •

• R •

• S •

• T •

• U •

• V •

• W •

• X •

• Y •

• Z •

COUNTERPOINTS

Studies in the Postmodern Theory of Education

General Editor
Shirley R. Steinberg

Counterpoints publishes the most compelling and imaginative books being written in education today. Grounded on the theoretical advances in criticalism, feminism, and postmodernism in the last two decades of the twentieth century, Counterpoints engages the meaning of these innovations in various forms of educational expression. Committed to the proposition that theoretical literature should be accessible to a variety of audiences, the series insists that its authors avoid esoteric and jargonistic languages that transform educational scholarship into an elite discourse for the initiated. Scholarly work matters only to the degree it affects consciousness and practice at multiple sites. Counterpoints' editorial policy is based on these principles and the ability of scholars to break new ground, to open new conversations, to go where educators have never gone before.

For additional information about this series or for the submission of manuscripts, please contact:

Shirley R. Steinberg
c/o Peter Lang Publishing, Inc.
29 Broadway, 18th floor
New York, New York 10006

To order other books in this series, please contact our Customer Service Department:

(800) 770-LANG (within the U.S.)
(212) 647-7706 (outside the U.S.)
(212) 647-7707 FAX

Or browse online by series:

www.peterlang.com

FSC
www.fsc.org
MIX
Papier | Fördert
gute Waldnutzung
FSC® C083411

Kolibri 360 GmbH
Ferdinand-Jühlke-Straße 7
99095 Erfurt, Deutschland
produktsicherheit@kolibri360.de